AF824749

Kindle Fire HDX and HD User's Guide Book: Unleash the Power of Your Tablet!

By Shelby Johnson

Disclaimer:

This eBook is an unofficial guide for using the Kindle Fire HD or Kindle Fire HDX and is not meant to replace any official documentation that came with the device. The information in this guide is meant as recommendations and suggestions, but the author bears no responsibility for any issues arising from improper use of the tablet. The owner of the device is responsible for taking all necessary precautions and measures with their tablet.

Kindle, Kindle Fire HD, Kindle Fire HDX and Amazon are trademarks of Amazon or its affiliates. All other trademarks are the property of their respective owners. The author and publishers of this book are not associated with any product or vendor mentioned in this book. Any Kindle Fire HD or Kindle Fire HDX screenshots are meant for educational purposes only.

Author Introduction

Hi, I'm Shelby Johnson, a technology enthusiast, Kindle Fire HD and new Kindle Fire HDX owner, and bestselling Amazon eBook author. With this newest tablet from Amazon, I absolutely love its many standard features and capabilities, but have found there is so much more that can be done with this amazing gadget than you may realize.

I've learned a lot of great things you can do with a Kindle Fire HDX and want to help others get more out of their Kindle Fire HD and HDX devices. I have created this guide to be the best way to get the most out of your Kindle Fire HDX. Without it, you may find yourself lost. At the very least, you will miss out on some of the amazing features this tablet offers that you do not know about.

This eBook is ESSENTIAL for any new Kindle Fire HDX owner. If you give a Kindle Fire HD or Kindle Fire HDX as a gift and do NOT purchase this guide, you're giving an incomplete gift.

Here is just some of the great info you'll find in this User's Guide Book:

How to use the new Mayday button for Kindle Fire HDX.

What new features Amazon included in the new Kindles.

Which features are missing from the new Kindle tablets.

How to use Second Screen.

How to use Kindle FreeTime to set parental control limits for children using your tablet.

How to get Google Chrome or other web browsers on your Kindle device.

How to use your Kindle Fire device as a phone.

How to sideload android apps onto your Kindle.

You'll learn all of the above and more in this book, which features color screenshots from the Kindle and step-by-step instructions on the processes involved, so that you can unleash the many powerful features your Kindle device offers!

For a listing of the best Kindle accessories see the Tech Media Source website.

Table of Contents

Introduction

With the popularity of tablet computers skyrocketing, Amazon has now launched its third generation Kindle Fire. These newest generations are the Kindle Fire HD and the Kindle Fire HDX. These devices comes in several versions, with a number of storage options for everyone from the novice, e-reader, and web surfer, to the professionally savvy social media moguls around the globe.

The Kindle Fire HDX is available in a seven or 8.9-inch screen or 8.9 inch 4G option, and the models range from 16GB to 64GB of storage capacity. With varying options and pricing depending on the model, it is safe to say no matter which one suits your fancy, you will be ready to take on the world in high definition style from the moment it leaves the box.

This eBook is perfect for novices, intermediates, and advanced techies looking for how-tos with their new device. For those who aren't necessarily tech savvy, but received the Amazon tablet as a gift, it can be a bit daunting to try to figure everything out. This eBook will help you get to know your tablet even better, while fully unlocking and unleashing many of the extra capabilities the device possesses!

The Kindle Fire

The new Kindle Fire HDX is chock full of technological genius that comes to life on a stunning 7" (also available in 8.9") screen that goes beyond high definition by providing exceptional color accuracy, reduced glare and amazing pixel density at 323ppi, 1920 x 1200. The accelerated processing is delivered courtesy of the Snapdragon 800, 2.2GHz quad core processor, which runs three times faster than the original tablets.

The Kindle Fire HDX adds to the quickness and unbelievable display by boasting 11 hours of battery life, in an effort to keep you up to speed on everything in your world without compromising the duration of its use, and even adds six more hours of life if you are only reading.

This ultra-light and incredibly durable device couples its speed and balance with amenities that include front and rear facing cameras (front on the 7", both on the 8.9"), personal online help using "Mayday", instant video downloads and the fun of X-Ray for your music files, so you can follow along to the words onscreen, or learn more about the band as you enjoy their songs. What's more is that it can be activated for movies, so you can interact with trivia or catch up on the backstories of characters.

With hundreds of new features in the Fire 3.0 Mojito operating system, including enhanced email and Second Screen, as well as carousel and grid views, you can personalize your new Kindle Fire HDX to match your every need, while it reflects your personality from the outside. Choose from different cases, screen protectors, keyboards, artwork and bedazzling effects as your personality dictates - or shuffle between several accordingly. This device is going to blow you away with its incredibly graphics and absolute functionality, so you are going to cozy up to it in no time, and want to care for it as if it will last a lifetime. As with all new and exciting devices, there is always more than meets the eye, so let's jump right into this device's offerings.

What's in the Box?

For starters, the new Kindle Fire HDX arrives in a beauty of a box, delivering an embossed appearance of the actual tablet on the front of the packaging to help drive home the excitement of what lies beneath its slickly designed confines.

Enclosed in the eco-friendly box is:

- The Kindle Fire HDX tablet
- One Pagc, two-sided "Getting to know your Kindle Fire HDX" Quick Card
- Square-shaped Fast Wall Charger (USB port on the top, plugs directly into the wall)
- Micro-USB Charge & Sync Cable

There is no hefty user guide, as the device works intuitively, and allows you to operate it functionally without confusion - or the fear that you have done something completely wrong. Should you become button happy, however, and find yourself stranded in technological limbo, help is just a tap away with the small on-board user guide and the Mayday feature (more on that later).

Getting to Know your Kindle

Once you unwrap the KFHDX, you will notice how sleek it is, and that it sits flush on a hard surface, so there is no rocking involved in its use, if you plan to sit it down instead of hold it. The back of the device contains a power button on one end of the device, and although it depends on how you are holding it as to which side it's on, if you can read the word "Amazon" correctly from left to right, than the power button will be in your right hand, and the volume buttons will be in your left hand (assuming you are holding it with two hands!)

There are two Dolby speakers on each side of the device, which are identified by their slatted design. Since the device is designed to sit up a bit - although flush - the speakers are visible when it is lying on a flat surface as well, so you do not miss out on any sound, no matter how you are using it.

On the very same side of the device as your volume control you will find a 3.5mm stereo jack port for your headphones. You can also use Bluetooth wireless headphones with this device, without an issue. In addition, on the side of the device that houses the power button lies a micro-USB connection, which is how you will power your KFHDX, and connect it to another source for data transfers when necessary.

Unlike the initial version of the Kindle Fire, neither one of the HDX options contain an HDMI connection. It has been completely removed from the device this time around. No worries, though - there are alternatives discussed in upcoming segments, should you rely on the HDMI for presentations and something fun like sharing videos on your television screen.

Getting Started with your Kindle Fire HD and HDX

The good news regarding your device is that is arrives with approximately 65% of its battery power intact, so you can start setting it up right away. Currently Amazon boasts that the device can be charged fully in fewer than six hours using the micro-USB connection that was delivered with the item, and stay charged for 11 hours with fully use as a result.

When you are ready to charge it, simply plug the micro-USB cable into the device, and directly into the wall charger, before plugging the charger itself into the wall. Since you should have more than enough battery power to get going, set aside a few minutes to set up your device's initial preferences, so you can start using it right away! Once you have completed the set-up, allow it to charge to 100% so you can test the battery life out for yourself. Although the device boasts 11 (or 12 on the 8.9" version) hours of use, it will differ for everyone. Also, it may take a week or two to start using it at a regular pace. Usually, when someone receives a new device, the battery will not last nearly as long as possible, because they are constantly using it in different capacities to learn the device's capabilities through and through.

Setting up the KFHD and HDX

Once you turn the device on, you are going to be ready to start setting it up with your personal framework and preferences, so you can enjoy its beauty without delay in a very personalized way. The first thing you will notice is that even when the device is locked the time (which is not correct yet, unless you got really lucky somehow!) and battery life with an icon and the actual percentage on display immediately.

Choosing Language

Once you slide the unlock slider that is also on display, you will be asked to pick a language. Currently the following options are available, and can be changed under your settings tab (more on this later!) should you (or someone else) accidentally change it - or set it up in a native tongue other than yours.

- Chinese
- English
- French and
- German
- Italian
- Japanese
- Portuguese
- Spanish
- UK English

Choose your language and click the continue button.

Connecting to Wireless Network

Your device will identify the closest WIFI network, and will require you to enter the password - assuming it is password protected (a lock symbol next to the router's name will tell you so).

Once you are connected to the WIFI source, it will immediately ask you to register your KFHDX using your Amazon account username and password.

Note: *If you do not have an Amazon account, you can tap "Create Account" and follow the onscreen instructions to create one. It will also automatically set the time for you, based on the router's current time.*

Note: *The Kindle Fire HDX tablet with 4G LTE connectivity will include an additional wireless selection to use your monthly data plan.*

Connecting Facebook and Twitter

The very first option you receive will be to set up you Facebook account on the device. If you currently have a Facebook account all you have to do is enter your username and password to continue. The same goes for Twitter. If you already have an account, simply enter your username and password and tap continue to be taken directly to your account.

If you do not have social media accounts, you can skip the process altogether, or set up an account at another time (or right then, if you prefer!). You can also setup one and not the other, if you have a preference, so there is no need to get a Facebook page just to honor the device's available options, if you are more of a Twitter person.

Should you decide to skip the setup and come back to it, simply do the following:

1. Swipe the screen downward from the top to reveal "Quick Settings."
2. Tap "Settings."
3. Tap "My Account."
4. Tap "Social Network Accounts" (choose Facebook or Twitter).
5. Enter the account information and tap "Done."

You can unlink your Facebook or Twitter account at any time going forward, and simply tap on their app icons to access your page directly whenever you would like.

Setting up Email

Next you are going to set up your email account - or a number of email addresses - directly on the device, so that you can keep in touch with friends, family or even the office. The ease lies in the fact that there is an actual email app on the Kindle Fire, so you can easily set up all of your important email accounts including Gmail, Yahoo!, AOL and Outlook, just to name a few.

Simply tap on the Email App Icon, and enter the email address of the account you want to add, and your password. Tap continue and allow it a few seconds to set up automatically. Repeat the process for each of the accounts you would like to add.

1. From the home screen, tap "Apps," and then tap "Email."
2. Enter your e-mail address, and then tap "Next."
3. Enter the password for your e-mail account, and then tap "Next."
4. If your e-mail account is not recognized, tap "Advanced Setup" to manually add your e-mail account to your Kindle Fire HD or HDX.

Change, add or delete email accounts at any time by using the steps listed above. There is no limit to the amount of accounts you can add, and each one will allow you to access them in their entirety including email threads, drafts, trash and complete folders. You will also be given the option to delete the email on your phone, but save it on your work computer, just so you do not miss a beat - or inadvertently overlook something.

Setting Up Calendar

When you are ready to manage your meetings, events, and day-to-day schedule with the Calendar App, you have a number of options to choose from so all of your happenings are listed in one place. You can sync Gmail, Yahoo! Mail, Exchange, Outlook, Hotmail, and Facebook calendars to your Kindle Fire HDX.

- From the Home Screen, tap "Apps," and then tap "Calendar."

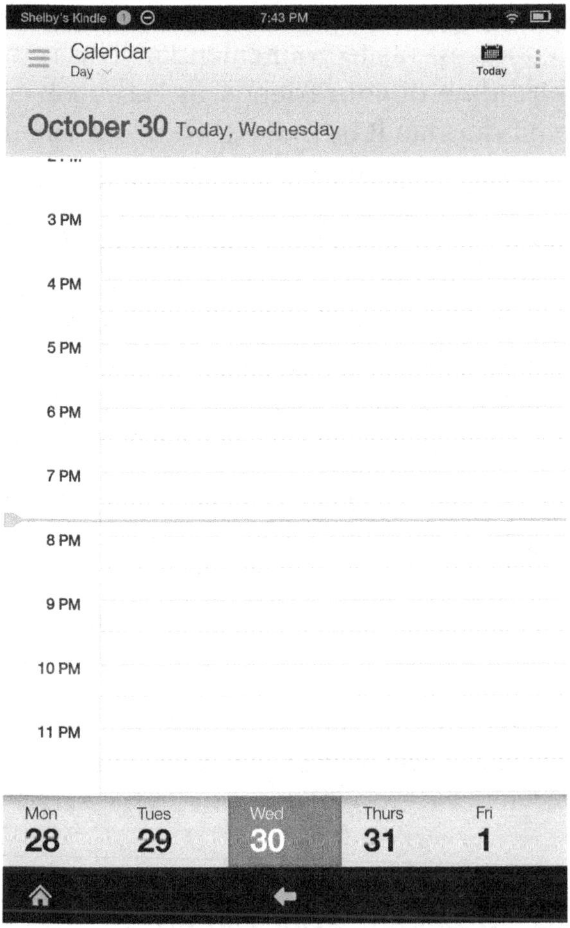

Once you are inside of the application you can change the calendar view to reveal the day, week, month or even lists you have created. You can also view or hide calendars just by swiping from the left side of the screen and tapping on any calendar that is synced with your Kindle. You can also view, create or edit and delete events you have created at any time

Add Facebook Events

If your Facebook account is linked to your Kindle Fire, you can add Facebook events to your calendar. This is incredibly helpful for birthdays of your friends, as Facebook is an awesome reminder that it is, indeed, your Aunt's birthday and you should acknowledge it right away.

Also, groups and invites are very popular on Facebook, so if you opt into an invitation to a party, it will place it on your calendar for you, and remind you when the date approaches. It certainly is easier than trying to remember all of it yourself!

To activate this feature, simply:

1. Swipe from the left edge of the screen, and then tap "Settings."
2. Tap "Calendar General Settings."
3. Select the option to "Sync Facebook Events."

When you add an event within your email accounts, it will automatically sync to your calendar app, so you will be on time and on schedule no matter where you are!

How to Deregister Your Kindle

If your Kindle Fire HD or HDX was a gift, or you purchased it with the help of a friend's account, you'll want to make sure it is not registered to them once you start making purchases with it. To deregister a Kindle Fire HD or HDX, simply select "Settings" from the home menu, and "Deregister" the device.

If you must deregister the device another way, logon to Amazon.com, and visit the "Manage Your Kindle" page. It will give you the option to deregister the device at any time. This is also important to remember if you ever decide to give away or sell your Kindle Fire HD to someone else since you won't want them making purchases on your Amazon account.

How to Change Your Kindle Name

1. Log in to your Amazon.com account on your computer or Kindle device.
2. Go to the "Your Account" drop down menu and choose "Manage Your Kindle."
3. On the left side of the menu chooses under "Your Kindle Account" choose "Manage Your Devices."
4. In this section you'll see all registered Kindles for your account. You'll see the name of your Kindle Fire HD or HDX, which you can click "Edit" next to and change the name of your device.

Note: *In this section you can also choose to opt out of sponsored ads (for a $15 charge) and Register or Deregister your Kindle device.*

Setting Up 1-Click Payment Option

The 1-Click payment option will be your payment method for purchasing content on your Kindle Fire HD or HDX. In order to choose the payment method or change it, you will need to visit your Kindle account at Amazon.com. Once there, click "Kindle Payment Settings" and then choose "Edit" to the right of "Billing Method."

Navigating the Kindle Fire

Chances are that by now you are pretty enamored with the screen's display; it is a thing of beauty. So as you begin to use your Kindle Fire HDX, there are a few tips to the navigation. Feel free to bounce around for a while to see which swiping motions and views work best for you. Remember, everyone has a different idea of what works for them technologically, and this device gives you the opportunity to embrace your personal approach to organization.

Initially your display will appear in a carousel, meaning all of your digital content will appear on screen, in an evolving line that you can swipe from right to left, or tap to read, watch or listen to. When you want to remove something from your carousel, simply tap and hold the item until you are given a list of options - one of which will be to remove it from the carousel. This does not delete it from your system, however, so no worries there.

Also, along the top of your screen you will see the following options, at all times:

- Search
- Shop
- Games
- Apps
- Books
- Music
- Videos
- Newsstand
- Audiobooks

- Web
- Photos
- Docs
- Offers

Once you are inside one of these libraries, you can swipe from the left-hand side of the screen to review content offers, recommendations and wish lists. You can also view different eBooks, music lists or movies without returning to the home screen to do so. This is a great way to remain focused in the category you are in, instead of jumping back and forth from home to music (or movies, or otherwise).

Swiping up on the carousel screen can access the home screen, which will contain ever-present content including your Silk Web Navigator, Email, Help, Camera, Calendar, Contacts and Shop Amazon icon. Simply put, place your finger on the device and push up and the home screen lies below.

Options Bar

Within any screen - other than the home screen - you can access the options bar, which allows you to go back to the previous page by pressing a back arrow, search for content using the magnifying glass icon, and the ability to get to the home screen by tapping the icon that looks like a house.

There is one more icon, which is comprised of three lines. This is the menu icon, and will provide various functions and settings; depending on which content library you are in (not pictured in the above screenshot).

Notifications and Settings Bar

As seen in the screenshot below, your Kindle name appears on the far left, with the time in the middle and your battery power level at the far right. Next to the battery power level indicator you will also notice the Wi-Fi signal strength indicator, and a Bluetooth indicator.

To access your notifications you will simply tap on the time and hold down, then drag down to reveal settings options as seen above (Auto-Rotate, Brightness, Wireless, Quiet Time, Mayday, and Settings) and your most recent notifications.

Note: *The above screenshot is from the Kindle Fire HDX. The Kindle Fire HD will have different settings options when you reveal the Notifications and Settings Bar. KFHD does not have the Mayday option.*

Choosing Carousel View (Most Recent) or Grid View (Favorites)

The fast and exciting operating system on the KFHDX allows you to view your content in two unique ways: Carousel or Grid view.

Since everyone is different, the preference is all yours! Fire OS 3.0 includes an enhanced, content-first home screen that allows users to easily switch between viewing their recent content and apps in Carousel View or favorites in Grid View.

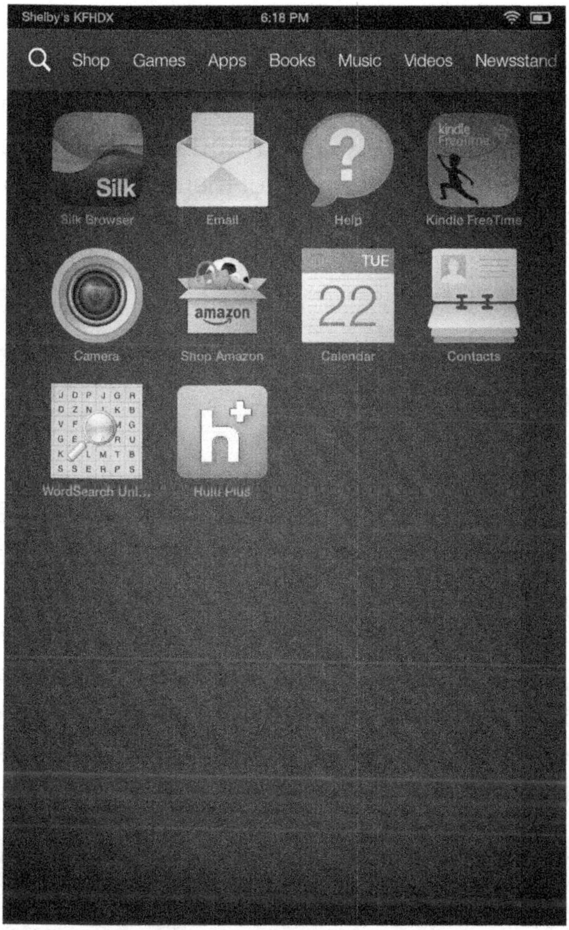

When you want to place an app, movie, book or other content into your favorites, simply tap and hold the item until you receive options. One of the options will be to move to favorites, which will automatically place it in a grid where the home screen located (or the next page or pages, depending on how much you have!).

When you want to get away from the carousel, simply swipe up to reveal the gridded version of all of your favorite things. If you prefer the carousel, which will show items that have been used most recently, keep in mind that it will bury items that have not been used in a while, including certain programming options like Hulu or Netflix.

Exploring Your Carousel

The carousel is the middle part of your home screen with content categories and the large icons below those. The heading will list the categories, and each of the components located within their selective media types. For instance, if you are in the category "Books," only the books will be on display. The same goes for music, apps, videos, newsstand items and audiobooks. You can click on the library option "Shop" to purchase new content at any time. You will need to set up a 1-click payment option in order to make these purchases.

With the large icons you see below that, your device will display any recent webpages you've visited, apps you've used, documents you've read, songs you've played or other recent items you have accessed on the Kindle Fire HD. You can use your finger to scroll through your current selections, and tap any of the items to open them or access them. You can also press down on these items to bring up a menu of options, including the option to remove an item from the carousel, from the device, or to add it to your favorites. Adding favorites will be covered more in-depth in a later section.

How to use the Kindle Fire HD and HDX

There are many things you will want to know how to do with your new Kindle Fire device. The following sections give detailed answers to your most comment "how to" questions.

How to Save Favorites

The Kindle Fire HD smartly allows you to save and access all of your favorite items so you can connect to them quickly, as seen in the image above. The favorites can be accessed by tapping on the star symbol on your display from most sections. You can add all of your favorites in just a few simple steps.

Carousel and Grid Items

The Carousel generally displays the most recent apps, webpages, documents, music, videos, and other items you have accessed on your device. You can add any of these to your favorites, which will appear in your Grid view. To do so, simply press down and hold on the carousel item.

Favorite Websites

Go to the web browser. Logon to the website you would like to add, or bookmark, as a favorite. Press your finger on the page itself until you receive a pop-up window with the following options: Close tab, Close other tabs, and Close all tabs, Add to bookmarks, Add to device favorites. Click "Add to bookmarks."

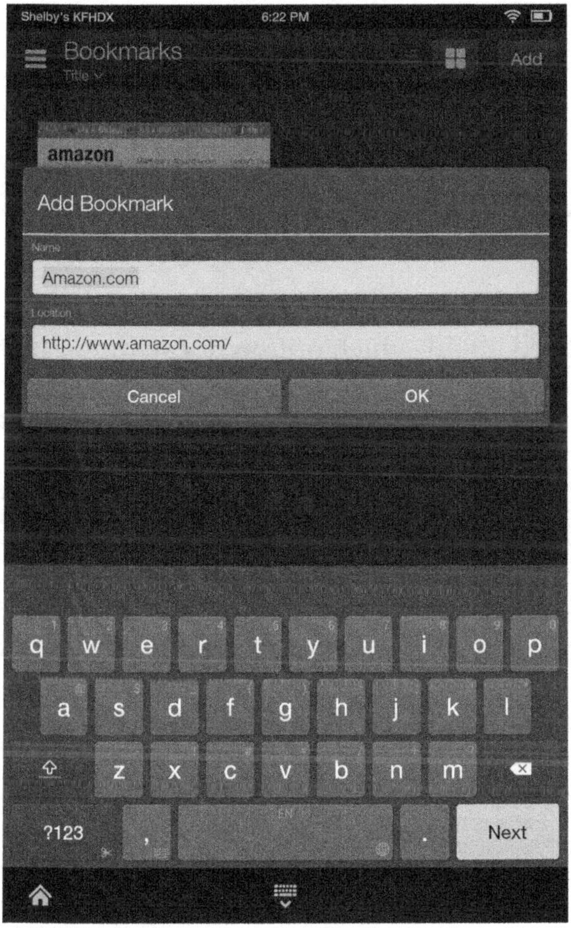

When you choose this option is will add the page to your device's bookmarks' library for easy access when surfing the Internet. If you choose "Add to device favorites" it will appear under the home page's star-shaped icon. This option is okay too, but you do not want to overcrowd the favorite's area with webpages, when they can be neatly filed under the designated "Bookmarks" section.

You can remove any of your favorites at any time by simply holding your finger on the icon or bookmarked page and choosing the "remove" option when prompted.

How to Adjust Kindle Fire HD Settings

With the amount of settings on any electronic device, it is easy to get confused about which option does what. The following explanations should help clear up any confusion going forward, so your device is used optimally at all times.

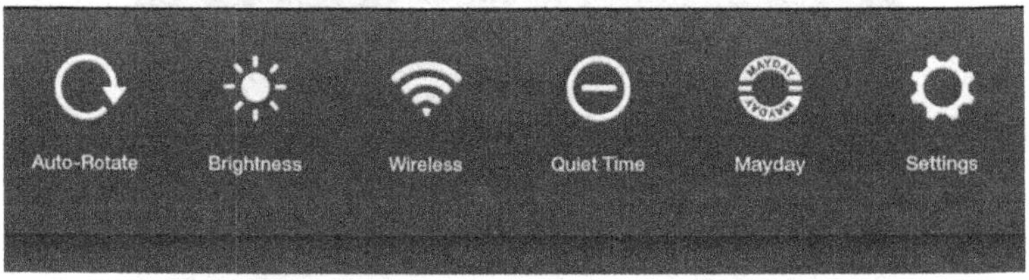

To access the various settings for your Kindle, hold down on the clock/time area at the top middle of your Kindle Fire HD screen, and then drag it down. You will reveal options such as "Auto-Rotate," "Brightness," "Wireless," "Quiet Time," "Mayday," and "Settings" option.

Note: *The above screenshot is from the Kindle Fire HDX, and the Kindle Fire HD will not have all of these options.*

Auto-Rotate

The Auto-Rotate feature is incredibly simple to use. Simply tap the "Auto-Rotate" button to turn the feature on or off. When it is on, your Kindle's screen will not automatically rotate when you change the device's orientation. When it is off, your Kindle's screen will rotate when you reorient the device.

Brightness

Tapping the "Brightness" button brings up the display brightness control. You can see the control in the screenshot below.

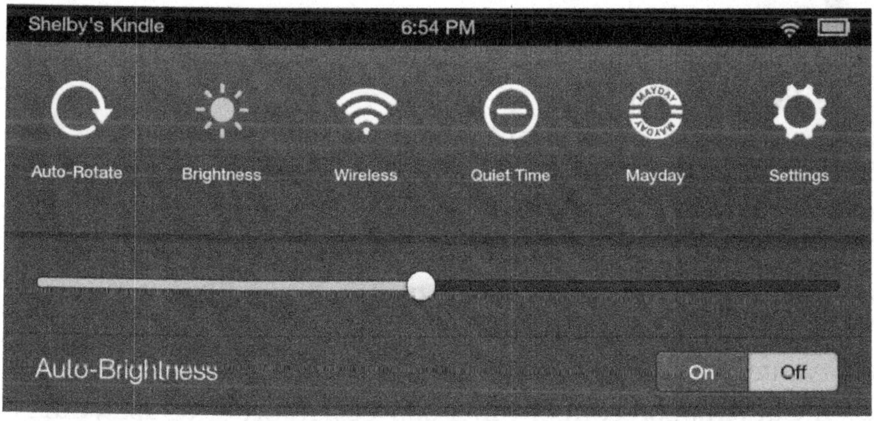

Simply move the brightness right or left to either brighten or darken the screen. You can also set "Auto-Brightness" to "On" to have the screen brighten automatically if you would like.

Wireless

This is where you will see any available Wireless network(s) and also where you can use Bluetooth. There is ale an Airplane Mode option here that will shut off all wireless connectivity on your device. This is helpful when traveling by airplane should you be instructed to shut off all wireless devices, but would like to still use your Kindle for other features.

How to Set Up and Use Wireless Connections

The following sections detail how to set up and use your wireless connections on your Kindle Fire HD.

What is Wi-Fi?

Simply put, Wi-Fi is the technology that allows your electronic devices to exchange data wirelessly over a computer network. This includes high-speed internet connections that are accessible via hotspots in public places. You will generally use Wi-Fi to surf the internet, e-mail friends, family, and colleagues, and access the vast array of multimedia content online at Amazon.com or other sources. In order to use your Kindle Fire HD with Wi-Fi, you must have a wireless router or network to access in your home, office, hotel, or other location.

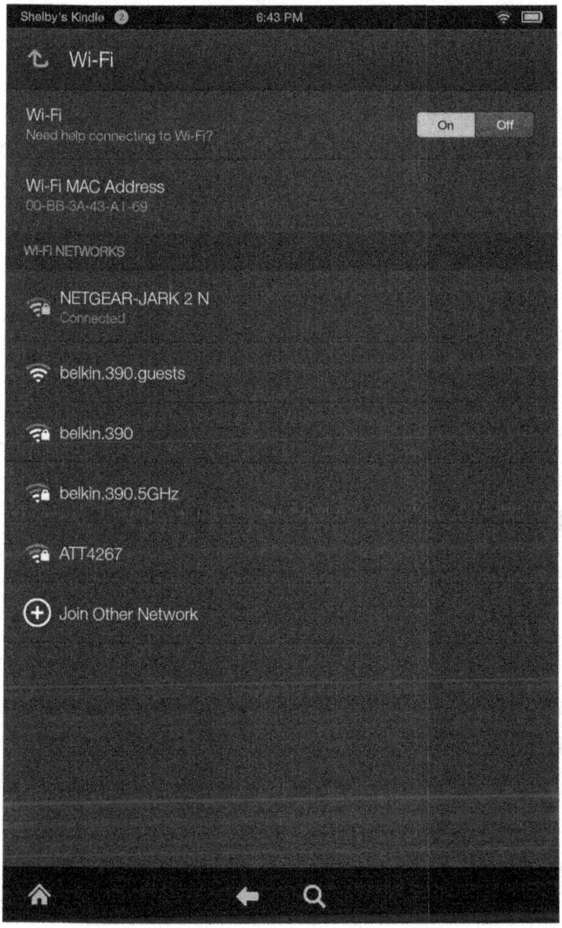

In some instances, such as staying in a hotel that offers free Wi-Fi service, you may need to obtain a username and password to be able to access the Wi-Fi on your Kindle Fire HD.

What is Bluetooth?

Bluetooth is an open wireless technology that exchanges data over short distances, using radio transmissions. It operates from a fixed or mobile device, in this case your Kindle Fire HD, to create a personal area network.

When Bluetooth is enabled on your device, you will see a small Bluetooth symbol up on the top area of your display, near your Wi-Fi symbol and battery status indicator.

This technology can be used for phones and wireless headsets or intercoms. In the case of the Kindle Fire HD, it can be connected to a wireless mouse, keyboard or printer. It can also allow movies and music to be played on wireless devices such as speakers or television screens with Bluetooth enabled technology.

One example of a way Bluetooth can be used is to transfer files such as pictures, music or videos from a smartphone to your Kindle Fire HD. To do this, the devices will need to be "paired" with one another. You must have a compatible device in order to do this that has Bluetooth connectivity. There are many smartphones, laptops, computers, and handheld devices that have Bluetooth capability.

How to Use Airplane Mode

The term "Airplane Mode" is derived from the ability to use your mobile electronic device without causing interference to other electronic signals around you. Simply put, when your Kindle Fire HD is in Airplane Mode, your ability to use the internet or Bluetooth is shut off. This is best used as one would guess, when traveling by airplane.

This means you can still use your Kindle Fire HD to read eBooks or documents, play games, listen to music, watch movies or use the camera of the device without interrupting the signal transmissions around you—say, on an airplane. This makes the device safe to use when it typically would not be permitted due to the possibility of signal interference. All of your activities on the device will be done offline, with Wi-Fi and Bluetooth inactive. If the plane you are traveling on offers Wi-Fi service, then you may be able to switch Airplane Mode off and use your tablet as you normally would.

Quiet Time

When you tap the "Quiet Time" button, it turns orange, and the "Quiet Time" symbol appears beside your Kindle's name. You can see this in the screenshot below.

Amazon created the quiet time button to allow you to perform tasks like reading an eBook without the interference of bothersome notifications. When you want your device to go completely quiet, simply press the button, and you will not receive notifications that will disrupt you.

Mayday

If you need any help with your Kindle Fire HDX, simply press the "Mayday" button. Amazon has live help available 24 hours a day, 365 days a year. Mayday calls are typically answered in just 15 seconds. Find out more about this amazing feature later in this book.

Note: *The Mayday button is only available on the Kindle Fire HDX models.*

Settings

Below are the settings available for the Kindle Fire when you tap on the "Settings" button.

How to Sync

The "Sync" button will allow you to sync all your content with other devices. For instance, you may have watched a movie or read a book on another Kindle device you have, and this button allows you to sync to your furthest location watched or read.

"Sync" will also make sure that all your recently purchased content is available for your use on the Kindle Fire HD.

Help & Feedback

This is a general section where you can get assistance to figure out troubleshooting or how to use specific Kindle settings and features. The on-board User Guide is located here, as well as the option to get customer service with your device. There's also a way to provide user feedback about the various features on the Kindle Fire HD, and send it to Amazon to help them improve for the future.

My Account

In this area you will see your Kindle Fire HD account info including whom the device is registered to and your kindle.com e-mail address. Also, you can Register or Deregister your Kindle, Manage Social Network Accounts, and Mange E-Mail, Contacts or Calendars on the device.

Applications

This area of your settings will allow for turning the various "Notification Settings" "On" or "Off" for all of your installed apps. Additionally, you can view all of your installed apps on the device, and Sync any Amazon content (eBooks, videos, music) you may have recently purchased elsewhere, say on a computer or laptop for example.

In addition, there is an "Amazon Applications" area here where you can tap on the setting for any of the following: Amazon GameCircle, Amazon Home Recommendations, Apps, Audiobooks, the E-mail, Contacts, Calendars section, Music, Silk browser, and Videos. Use these to adjust settings for any of those particular content or app functions on your device.

For example, going to the "Silk" option will allow you to set up various features and settings on the standard Silk browser. These include blocking pop-up windows, or setting a specific Search engine such as Google on your Silk web browser.

Device

This option on settings will give you basic details about your device and include a few settings you can adjust. Here's a bit about each of the Device listings.

About - this section gives you various technical details including the latest version of firmware installed on your device. It will also give you the option to "Update Your Kindle" if a new version of firmware is needed. Other info in the "About" area includes the device serial number, Wi-Fi MAC address, and Bluetooth Mac address, which are generally needed to resolve certain issues.

Storage - Tap on this to learn how much of your Kindle Fire HD storage you are currently using on the tablet, and how much is still available to use. You will be able to see a breakdown of your eBooks &Newsstand items, Audiobooks, Music, Video, Photos, Docs, Personal Videos, Apps & Games, and Others. Use this to gauge where you might free up extra space should your device be getting low.

Battery- This is an indicator bar of how much power is left on your Kindle Fire HD. It will also tell you if the device is charging when plugged in to a charger or USB slot. There are some apps and ways to preserve battery power covered later in this guide.

Date and Time - Use to set your current time zone, should you be traveling. You can also choose a 24-hour time format if you would like.

Allow Installation of Applications -This option will be covered later on, but it will allow for installation of certain apps from other app stores besides Amazon's store.

Reset to Factory Defaults - This may be a last resort setting you come to should your item have a major issue or you are getting ready to sell or give someone else the device. It will remove all personal data from the Kindle Fire HD.

Location-based Services

Certain apps will use your location to assist you better. Examples might be a directions or map program that needs to know your current location, or a weather app that tells you your area's current forecast. Amazon has made it clear that some third party app companies may collect this information too; so make sure to tread with caution when installing certain apps that may not seem reliable. Often times, customer reviews will tell the story on the app as well.

When going to a particular app in the Amazon Appstore, you will see it may have "Permissions" which are listed. This section of the app's details should tell you if it will use location information.

You can easily shut off this feature as well, but choosing "Off" at the "Location-based Services" area on the settings menu.

Language & Keyboard

Use this to set the preferred language for your Kindle Fire HD. At this time the choices include Dutch, English (UK or US versions), French, Italian, and Spanish.

The Keyboard settings area gives you a wide range of options, including having sound when you tap on the keys of the keyboard, Auto-capitalization, Auto-correction of spelling errors, and Correction or Spelling Suggestions. These last two options are helpful should you misspell or mistype a word, as it will suggest the words it believes you meant to type, or a correct spelling of one you misspelled.

Security

With these settings you can restrict access to your Kindle Fire HD through the use of a passcode or PIN, as well as through the use of a Virtual Private Network.

How to Set Security Options for Your Kindle Fire Device

Among the More options is "Security." This setting is particularly helpful for protecting your device from unauthorized use and safeguarding any important data you have on the device. One of the best things you can do is to "Lock Screen Password" as described next.

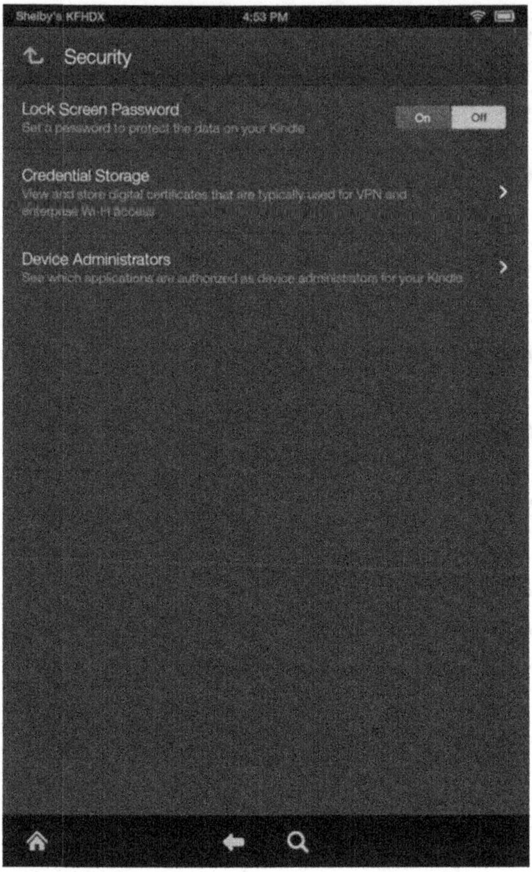

You can use this option to set up a password for your Kindle Fire HD. This way when the device is locked and someone goes to unlock it, they will need to enter your PIN # to use the device. This is helpful to protect your device in the event it is misplaced or stolen, so that any vital data is not stolen. To set up a Lock Screen Password, touch this selection on the screen, and then enter a PIN # of at least 4 characters. Make sure it is a number you will remember, but not one a stranger would easily guess.

The VPN option will allow you to use the Kindle in a virtual private network, or VPN. To do this you will need to download a special app from the Amazon Appstore. This is a more complex setup that may require the help of an IT professional from your office, or someone in your home with knowledge of your network.

You can also set "Device Administrators" in the Security area. This will allow multiple users to be in charge of the Kindle Fire HD, such as a set of parents, teachers, or others. It can be helpful for keeping the device protected from unauthorized use by children, for example.

How to Pair a Bluetooth Device with Kindle Fire HD and HDX

1. Drag down the notifications bar at the top of your display (press and drag down where the time shows in the top middle of the screen).
2. Tap on "More," then tap on "Wireless."
3. Tap on "Bluetooth."
4. Make sure to tap on "Enable Bluetooth" so it is on.
5. The Kindle Fire HD will search for any devices in the region with Bluetooth connectivity turned on. It may take several minutes to find some devices.
6. Tap on the device that you want to pair to begin the pairing process.
7. You will receive a pop-up box on your computer or other device, as well as a pop-up box on the Kindle Fire HD explaining how to complete the pairing process. Generally, this involves checking both devices to make

sure a numerical code matches on them. In some cases, you will need to type the code in for one of the devices.

8. Once paired, you can use your computer or other device to send files via Bluetooth to the Kindle Fire HD, as long as both devices remain powered on and the Bluetooth connectivity remains on as well. Some examples of devices that I've personally used this with successfully are LG smartphones and a Mac Mini computer, but there are many other items out there that work.
9. Some of the files you can send via Bluetooth to your Kindle Fire HD include MP3's, documents, photos and videos. Once they have been sent to your Kindle Fire HD, they'll show up in the notifications area or in the appropriate section of your Kindle Fire HD device (i.e. photos, music, docs, etc.).

Note: *Keep in mind that some devices may show up on the Kindle Fire HD, but might not be Bluetooth compatible. For example, the Apple iPhone 5 is among the non-compatible devices. Also, pairing a keyboard or headset will require a different process as described in that device's instructions.*

How to Purchase Content for Kindle Fire HD and HDX

You can purchase content, including music, movies, books, magazines and television programming through the "Shop" app on your device. Remember, you will have to have your One-click purchasing account set up before you do so, but that takes only a few seconds.

To make purchases, you can tap on the "Shop" option at the top bar of your Kindle Fire HD, or tap on any of the various types of content (i.e. Games, Apps, Books, Music, Videos). Tapping on these choices will bring up whatever content you currently have stored on your device and in the cloud. It will also have a "Store" option you can tap on to start shopping for more content.

In addition to purchasing items that will exist on your tablet, you can purchase ANYTHING that is available at Amazon.com. If you would like to buy a sweater for your mother for her birthday, simply browse the available selection, purchase the item, and ship it to her directly! Do you want to buy a bicycle for your son or daughter's birthday? Find the one you want at Amazon, and purchase it with one click. Their inventory of goods and technology is endless, and so is your ability to purchase them using your tablet.

Amazon Prime Service

Subscribing to the Amazon Prime Service is easy, affordable, and provides members with an endless amount of perks, simply for signing up and paying the annual fee. Amazon members can go sign-up now for a one-month free trial to start enjoying unlimited instant streaming of thousands of movies and television shows, thanks to Prime Instant Videos. In addition, you will receive one free eBook borrowing opportunity each month from the Kindle Owners' Lending Library.

Also, going forward, you will receive free two day shipping on virtually every item you order from Amazon, with no minimum order size! After the first free month, the subscription is available for $79 per year, which will pay for itself in no time.

Many Kindle Fire HD owners and Amazon members opt for this service not only for the shipping, but also as an alternative to subscribing to Netflix for their streaming movies and TV shows, due to the wide selection of content Amazon offers.

How to Transfer Files via USB on a PC

You can transfer photos, music, documents, video, and many other file types from your computer to your Kindle Fire HD, and vice versa. You'll need to connect the included USB cable from your Kindle Fire HD to your computer first.

For PC's it should automatically connect your Kindle Fire HD to your computer and it will show up as a drive you can open. Once you've done that, you can simply drag and drop files from your computer to the Kindle Fire HD, and vice versa.

For some computers and MAC's, you will most likely need to go to the kindle.com/support/downloads webpage and download a special USB Transfer Tool.

How to Transfer Files via USB on a MAC

For a MAC computer, you'll need to use the free Android File Transfer app to complete USB transfers. Go to Android File Transfer App < http://www.android.com/filetransfer/> in your computer browser and follow the onscreen instructions to download this utility.

Once you have it installed you can connect your Kindle Fire HD via the USB cable to your MAC. The Android File Transfer utility will show as a drive on your computer, similar to any other USB drive you might plug in. Once you are able to connect your Kindle Fire HD to your computer, you can drag and drop files between your Kindle and MAC computer. Make sure to click on the eject button next to the Android File Transfer drive on your MAC before unplugging the USB cable, just to be safe.

How to Listen to the Radio on Your Kindle

Did you know it is possible to stream local and national radio stations on your Kindle Fire HD? There is a great app called Tune In Radio, which is free at the Amazon Appstore. With this app you can listen to a wide range of local radio stations over your wireless connection. Additionally, you can choose from different categories including music, sports, news, talk and podcasts. The radio app will play as you use various features on your Kindle Fire HD making it a great way to enjoy music while using the tablet device.

How to Choose a Kindle Fire HD or HDX Web Browser

The Kindle Fire HD comes standard with the Amazon Silk browser, and the verdict is still out from most users as to whether this is the best option for your tablet. There are a few alternate browsers you might consider such as Dolphin Browser, or Google Chrome, which you may find to offer a more appealing and comfortable web browsing experience.

Dolphin Browser vs. Amazon Silk

Dolphin Browser is a quick moving, smart and free web-based browser that delivers stunning graphics, combined with impeccable search results, and will allow you to open your favorite pages with the tap of your fingers. PC Magazine calls Dolphin the most capable browser available for your tablet, and over 50 million downloaders could not agree more. Many Kindle Fire owners opt for this browser to help with resizing webpages to their liking for easier display and readability.

Amazon's Silk browser comes standard with the device. It basically chooses between two web browsing research options, with one of them being Amazon's servers. This means the information provided might be centered on Amazon's recommendations, instead of fully disclosed results.

Many users decide to install the Dolphin Browser, giving them another option to browse the internet and better use the capabilities of the World Wide Web on their tablet. The biggest pro to obtaining the browser is it will allow for the use of Adobe Flash, which is a valuable component of many website. Adobe Flash support on the Dolphin browser allows for viewing of website video content on your Fire HD.

How to Install the Dolphin Browser

On your Kindle Fire HD, go to "Settings," then "Device," and Enable "installation from unknown sources." Bypass the warning message on the popup box by tapping "OK" to enable this setting.

1. On your Kindle Fire HD, go to "Web" and then search for "Dolphin Browser HD 8.5.1."
2. Use the site you found, or you can try the one located at the XDA-Developers forum. (Remember, you are using these sites and links at your own risk.)
3. Download the "APK" file for 8.5.1 you find at the website.
4. Access your Notifications. Find the recent download. Tap on it to install.

Now you have the Dolphin Browser as an alternate web browser with many other settings you can set up.

How to Install Google Chrome Browser

Another web browser that some users might prefer to have on their Kindle Fire HD is Google Chrome. Here's how to install the Google Chrome browser on a Kindle Fire HD. (Note: This involves "sideloading" an app which is only legal when you already own the app files yourself, for example from purchasing/downloading the app for your phone or other device)

1. Pull down the notifications/setting bar at the top of your display (tap where the clock time shows at the top middle of screen and drag down).
2. Go to "More," then "Device," then tap "On" for "Allow Installation of Applications," if it is not already on.
3. A warning box will pop up. Tap on "OK" to bypass the warning box.
4. On your Kindle Fire HD or HDX, go to the web browser and do a search for the following file "com.android-chrome-2-apk" (without quotes). There will be various forums and websites that are hosting this file. Be cautious as some sites may have spammy links or malware, but for the most part, forums such as the XDA Developers forum have been generally reliable for downloads of .apk files.
5. Download the file mentioned above to your Kindle. You will do this at the forum link above by tapping on the link provided. The file will start to automatically download.
6. Once the download has completed go back to your Home screen, drag down the notifications bar and tap on the recently downloaded .apk file.
7. You'll be asked, "Do you want to install this application?" Tap on "Install" at the bottom right area

of the screen. This will install Chrome Beta on your Kindle Fire HD.

8. Once the installation is complete, tap on "Open" on the bottom of the screen.
9. You will now be able to use Google Chrome as an alternate browser on your Kindle Fire HD. Once you first open it, you may be prompted to accept a user agreement, as well as sign in to your Google account. Signing into a Google account is not required to use the browser.

Note: *You may experience difficulty trying to view YouTube videos on the YouTube website with Chrome browser on your Kindle Fire HD.*

How to Install Adobe Flash on Kindle Fire HD and HDX

Note: *Flash isn't a necessity to watch videos on the Silk browser. Websites that display videos such as ESPN.com and YouTube.com will both work without it. In addition, Amazon has an "experimental streaming viewer" to watch videos from other sites that use Flash, such as NBC.com, ABC.go.com, or others. However, if you are unable to stream something, you may still want to try adding Flash capability.*

1. Go to "Settings," then "Device," then tap on "Allow Installation of Applications from unknown sources."
2. Download the Adobe Flash player from the XDA Developers forum.
3. If you haven't already, download a Dolphin Browser HD APK file (one that you have acquired through an app store).
4. Install ES File Explorer if you haven't already. Go to ES File Explorer App on Your Kindle Fire HD and Tap on it to display folders on your device.
5. Tap on the "Download" folder you see.
6. Tap the Adobe Flash icon and then tap "Yes" to install the Flash app.
7. Go Back into ES File Explorer, Downloads folder.
8. Tap the Dolphin Browser Icon and tap "Yes" to install.

Once you've installed Dolphin Browser, you can use it as one of your apps as another web browser which has Flash enabled. In the Dolphin Browser "Favorite Settings" you will also be able to keep Flash Support "Always On," "On Demand," or turn it "Off" completely.

How to Watch Movies on Your Kindle Fire HD and HDX

The Kindle Fire HD is a great device for entertainment, as it connects you to the vast selection of movies available for rental and purchase from the Amazon website. You can choose a new or older movie or TV show to watch, purchase it, and it will stream to your Kindle Fire HD tablet.

You'll be able to watch movies, TV shows, play games, use apps, and even browse the web on an even bigger and more vivid display, unleashing the Dolby sound and Hi-definition features!

How to Watch .WMV & .MOV Files on Your Kindle

Both .wmv and .mov files are a popular format for movies and videos. If you add them to your device, they will not show up on your device in the "Videos" Library. However, using the ES File Explorer app, you can view these videos on your device.

Once the app is installed, connect your Kindle Fire HD or HDX via the micro USB to your laptop. Simply drag the .wmv file(s) from your computer files over to your Kindle Fire HD's folders.

You can now open up ES File Explorer app on your tablet, and then find the .wmv file you want to view. Tap on it, and the video will begin to play on your tablet.

Note: For .mov files, you will be able to do the same as above, but you may receive a pop-up prompt asking what you want to view the video with. Tap on "ES Media Player." You might also want to tap on the checkbox "Set as the default app" before you do.

How to Use a Google Chromecast to Stream Video

The Google Chromecast ($35 at time of this publication) is another option for streaming video content to your TV. Chromecast is a "dongle" which Google released that you can plug into an HDMI spot on your TV for streaming content. Video content can be streamed from Netflix, Hulu Plus, YouTube, and in the case of some devices, anything from a Google Chrome web browser tab. While the device doesn't come compatible with the Kindle Fire HDX, it can be worked around through sideloading to enable it to stream content from certain apps. *(Keep in mind, .apk file sideloading is considered illegal, unless they are files you have purchased and own for an Android device.)*

1. Install the free Hulu Plus app from the Amazon app store.

2. Install a Google Chromecast app apk file (do this by adding the .apk file to your Kindle's files and then open it with File Explorer app).

3. Install a YouTube Android app apk file (open with the File Explorer App).

The Kindle Fire HDX was tested to see if it would work with the Chromecast, and if you have the .apk files listed above, you can add them to your Kindle Fire HDX folders and file directory. Once they're on the tablet, use something like the free ES File Explorer app to open the .apk files and install them.

Open the Chromecast app from inside ES File Explorer after you've installed. Go through the set up process to discover your Chromecast on your wireless network and make sure your Kindle Fire HDX discovers it.

Launch the Hulu Plus app, or either of the other two apps mentioned above that you sideloaded, and you'll see the Chromecast "symbol" on the video content you play on your tablet. Tap on that symbol and you can wirelessly stream to your Chromecast from the KFHDX. The same should work for the sideloaded YouTube app.

As of this publication, this method was tested with Hulu Plus and YouTube apps. It worked well in both situations, and has yet to be tested with Netflix apps of any type. Chromecast streaming did not work with the Amazon Instant video content that was tested.

Amazon Prime vs. HuluPlus vs. Netflix

While there are thousands of movies at Amazon you can buy or rent, there is also the aforementioned Amazon Prime service. If you have this subscription based service, you can choose from any movies or shows listed in the Prime Instant Video area and watch them completely free of charge!

There are other options for watching movies or TV shows, two of which come from the subscription-based apps Netflix and HuluPlus, mentioned later on.

As of this report, Amazon Prime is $79 a year, with the first month free for anyone who buys a new Kindle Fire HD. HuluPlus costs $7.99 a month, with cancellation available at any time. Netflix ranges in cost, with streaming available for new members for $7.99 a month. Current DVD/Blu-ray subscriber accounts can add a streaming plan for $4.99 a month for limited streaming, and $7.99 for unlimited streaming to multiple devices.

When comparing these costs, Amazon's service is about $16 cheaper than HuluPlus or Netflix and adds the free 2-day shipping feature for many products on the website, which is why many consumers go for their service. The shipping charges saved alone can make sense for many individuals who shop online often.

It is best for anyone who wants to subscribe to one or several of these services to check out that particular provider's content selection to determine if the monthly cost justifies the content you will be able to stream and watch.

Two other options you might consider are HBO GO, which is a streaming video service available to HBO cable or satellite package subscribers, and Vudu.com. Vudu is another online movie rental and purchase site which offers standard and hi-definition movies you can rent or download. I tested out Vudu and it works well with the Dolphin Browser and Adobe Flash installed on a Kindle Fire HD. Vudu does not work on the Silk Browser because it does not have Flash.

How to Watch YouTube Videos on Your Kindle

Depending on which web browsers or apps you have installed, you can view YouTube videos on your Kindle Fire HD. YouTube videos should play on the standard Silk browser that your Fire HD uses for web browsing.

YouTube videos tend to look excellent on the Dolphin Browser that can be installed onto your tablet. That is covered in the "How to Install Dolphin Browser" section of this book. Flash may be required, and installation of that is also covered in the book. You can also "sideload" a YouTube android app to use on your device, although this is not recommended, since YouTube videos will play through the Silk browser.

A test of Google Chrome browser on Kindle Fire HD revealed some issues playing videos on the YouTube website via the tablet.

A great free app you can use is the TubeMate app described in the next section. Not only will you be able to watch YouTube videos with it, but you can also download them onto your device to watch at a later time. See more information on how to do this in the next section.

How to Download YouTube Videos onto Your Kindle

While Amazon offers a lot of great content in terms of videos, you may want to really stock up on videos, say if you are going on a trip and need something to stay entertained while traveling.

To download YouTube videos to the Kindle Fire HD, you can use an app called TubeMate, which is not found at Amazon Appstore. Here's how to install and use the app:

1. Go to the Web browser on your Kindle Fire HD and enter tubemate.net in the bar at the top.
2. Once at TubeMate.net, tap on "Download (Handster)" button.
3. At the next screen hold down on the blue "Download" button you see. You may need to enlarge your web screen display to make the button larger, and then press down on it.
4. A box will pop up. Tap on "Open." The Tubemate download will begin.
5. Once completed, the Tubemate app will be in your notifications area. Tap on the TubeMate download there, and then tap on "Install" at the bottom of the next screen.
6. Tap on "Open" button at the bottom of next screen to open TubeMate.
7. When a pop-up box appears, you can tap on the checkbox next to "I don't want to see it anymore" and then tap the "Close" button.
8. You'll now be able to search for YouTube videos to watch. Once they're playing you can tap on the green

arrow at the bottom of your screen to download the video to your device. Videos can be downloaded as 1280x720 since the Kindle Fire HD is a "high-end device."

9. The video will be downloaded and will show up in your notifications area. You can also see all the videos you download from the TubeMate app while in TubeMate. Just click on the folder icon at the lower right-hand area of the screen.

There's much more you can do with the app in terms of settings, but these are the basics to download YouTube videos for future viewing on your Kindle Fire HD.

How to Take a Screenshot

Here's a neat trick with your Kindle Fire HD and HDX. You can easily take a screenshot photo of whatever is currently on your device screen. This may be helpful for showing others a particular app or settings area of your device, or many other reasons for showing off the screen. Here's how to do it:

1. Make sure you have the particular screen displaying on your Kindle Fire HD that you want to take a screenshot of.
2. Hold down the Power and Volume down buttons.
3. The screen will flash and make a sound indicating a picture was taken. The screenshot will briefly show up on your screen, and then will be stored in your "Photos" area of the device.

You can then move these to another device via the USB connection to a computer or laptop, and upload them or store them in your Cloud if you'd prefer.

New Features for KFHDX

There are plenty of new features in store for Kindle Fire HDX users, which will make their lives easier by the second! Whether you are looking for faster movement, exceptional graphics or flawless business integration, you will find everything you are looking for with this device. In fact, mid-November marks a highly anticipated update of the operating system that will supply even more capabilities and effortless use, allowing you to enjoy your device as if it is new all over again! First things first, the camera!

Camera Options

If you have the 7" version, your device will only be equipped with the high definition, front facing camera, which is great for Skyping with friends or having meetings online.

If you have the 8.9" version, you will have an eight-megapixel camera on the backside as well, which is perfect for taking pictures at a moment's notice. Either way, there are a number of things you can do with the device, and the pictures taken or housed on it. If you would like to add a library to your new device, download existing albums from Facebook, or add photos and personal videos to your Amazon Cloud Drive from your PC, Mac, or phone and they will sync automatically with your device.

Camera App/Photo Roll

The camera app on the KFHDX allows you to edit images directly on the screen, simply by selecting a photo you want to alter, and then using the tools available to remove red eye, crop the image as a whole or change the look of the image to black and white or sepia tones.

Since only one version has a rear facing camera, taking pictures may be a little tougher with the 7" version, but it makes for a perfect tool for video conferencing, which is great. You can also edit images you did not take with the device, which can be great for future use. You may also delete images the second you take them, or roll through the collection to see what can stay and what can go.

The beauty of the device is its memory capacity, so if you are in a zone, start taking pictures like crazy and analyze them later to determine which ones stay - and become part of the collection - and which ones go.

Using New Mayday Button for Live Help

The Mayday customer service feature is only available on the Kindle Fire HDX devices, and not the previous iterations carrying the "Fire" name. This awesome addition allows you - the user - to tap a button and connect with an Amazon representative 24 hours a day, seven days a week, and 365 days a year.

This exceptional feature will connect you with a live person who appears on-screen and can walk you through any questions you might have about the device.

He or she will have the ability to take remote control of your device, by highlighting items on the screen, so you know where to look (now and) the next time you are stuck in a conundrum regarding the device's capabilities - or where something is located.

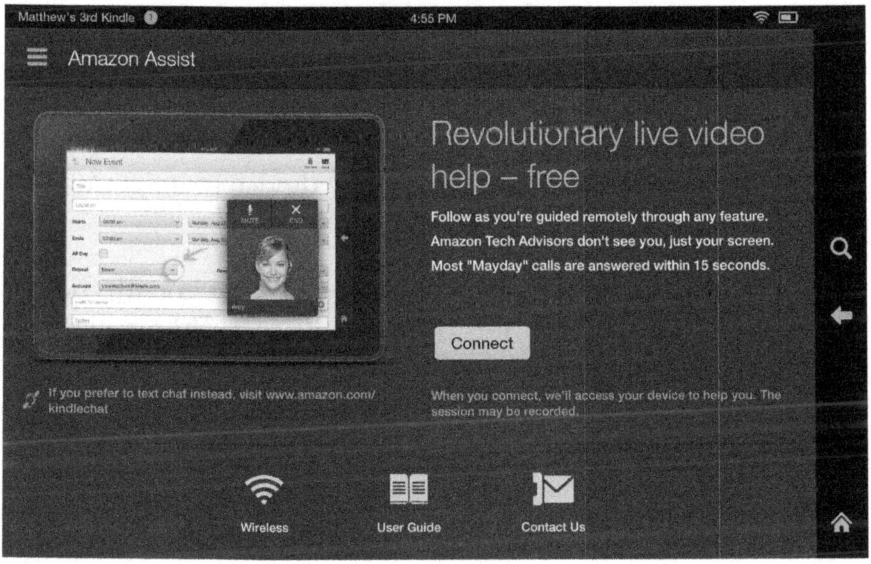

The availability of a Mayday representative is great because you do not have to sit on hold waiting for the right answer, and you will have a physical presence with you to show you through the problem without issue. So there is no more searching for a customer service representative varying description of an item on (or off) screen, and there are no more technical terms necessary to make you feel like a fool for owning such an awesome device.

This option is available at all times, so tap the feature and allow someone at Amazon to help you right away by literally pointing at the problem - or the solution.

Integrating Office & Productivity into Your Tablet

Amazon has added a great group of enterprise-friendly features, including the ability to connect to a corporate virtual private network (VPN) and the ability to centrally manage your Kindle Fire HDX tablet via mobile device management (MDM) solutions.

These options are perfect for someone who travels with their tablet, and wants to grab files from the office for home use - without having to lug their laptop around. With support for encryption of the device to secure data and support for Kerberos authentication, corporate users can browse secure intranet websites and connect effortlessly to download data files as needed.

It also allows you to separate your contents, so work items remain in that category, as do your personal items. You do not want your bosses snooping through your online dating profile simply because you have accessed the business network, and the KFHDX will ensure that is the case, by protecting your personal items from anyone else's remote view.

Reading and managing documents is simple, as you can decide to email them, sync them from a computer with Cloud Drive, clip them from the web directly from the web with the "Send to Kindle" option, or transfer them using a USB. Clipping articles from the web has been made easier, as your Kindle has a special area in the "Docs" option that allows you to send install links to your email address. The install links will give you add-ons for your web browsers, Firefox and Google Chrome, so that you can start clipping things and they'll be synched to your device for viewing on the go!

Kindle Fire tablets provide a robust exchange email experience with ActiveSync, so you can check and reply to work e-mail on the go, without accidentally tapping your personal account along the way. View Word, Excel, and PowerPoint files flawlessly - while editing, creating or saving ones of your own - and enjoy business apps that allow you to move around in a professional capacity so you can provide feedback to coworkers, clients and vendors in a timely manner, all while enjoying the entertainment aspects of your device as well.

Goodreads

Speaking of entertainment, Goodreads allows you to join over 20 million other readers - from strangers to best friends and acquaintances to virtual "friends" who share the same content preferences as you. You can see what everyone is reading, and share what you are reading along with adding your own personal reviews, building a library of favorites or contracting free material though the virtual libraries of others.

Share highlights and rate the books you read with Goodreads on Kindle, which is now an actual subsidiary of Amazon - purchased in March 2013. The activity will be flawless, and you will have access to one of the largest online book clubs compiled!

With over 12 million books to choose from, and thousands on the "free" list, you can read anything and everything that appeals to you using the reviews and ratings that appear alongside the title. You can also share the book's information and your review through your social media outlets. This app also has a barcode scanner, so you will never forget the name of a book, or its author, again. Simply scan the book's ISBN number, and it will add the title to your intended reading list automatically!

Printing from KFHDX

The Kindle Fire HDX tablet supports printing documents, spreadsheets, presentations, photos, and emails to your home or office's wireless printer without any hassle. Simply choose the document, image or email you would like to print, and select "Print" from the menu button that is available at the top of the document. The printer must be configured as an option - or at least obtainable through your Bluetooth capabilities.

Creating Collections

Depending on which device you purchase, you will be allotted a certain amount of memory on your hard drive: 16GB, 32GB or 64GB. With that space you can download apps, music, movies, eBooks, documents and peripherals in abundance, leaving them directly on the device for easy access. Keep in mind that if they are stored on your device, and not syncing through your Cloud Collections, that is going to be the only place you can find that data, which is perfectly fine should you want to keep it that way.

You are also able to take advantage of Amazon's amazing Cloud capabilities, so the space on your device is not devoured by entertainment. Although the amount of memory used on the device should not affect its operation - as the dual core processor is still lightning fast - it does make sense to keep your items in the Cloud - specifically for back-up purposes. This way, if anything happens to your KFHDX, you have everything stored remotely so you can download it again with ease, or at the very access it in an emergency.

Everything you purchase from Amazon is automatically stored in their Cloud, and can be accessed from any device, at any time, as long as you have an internet connection. You can also organize your content library into Collections like "Favorite Books" and "Sports Apps" that are synchronized with your other Kindle devices and reading apps, so you will always be on the same page - no matter where you are accessing your material from.

What's more is that the new KFHDX has a new1-Tap Archive. This option frees up space on your Kindle Fire HDX by identifying items that have not been used recently. Those items will be quickly moved to your Cloud for later retrieval, simply by tapping the 1-tap option.

At all times, no matter what you are downloading to your device, the KFHDX ensures that is it is transferred through optimization download, so your current use of the device is not affected by the activity happening in the background.

Downloading PRIME Instant Videos

First things first: When you purchase a new KFHDX, you will automatically receive a free 30-day trial of Amazon Prime. This means you can order anything you desire from Amazon, and ship it two-day without any extra cost. In addition, you receive access to the Amazon Prime Instant videos, books, music, and more simply by being a member!

Prime members enjoy commercial-free, unlimited streaming of thousands of popular movies and TV shows at no extra cost. Simply find the movie or series you would like to view, and tap on its icon to begin the download.

In addition, you can borrow any title from the Kindle Owners' Lending Library for free, and read one free book a month - or as they desire thereafter - without the pressure of due dates, late fees or charges of any kind. There are literally hundreds of different titles waiting to be checked out for free using this service.

Finally - and the coolest feature by far - is that KFHDX owners can download Prime Instant Video movies and television shows and watch them offline! This means you do not have to have internet access to stream your favorite release on an airplane, train or even in your car. Prime Instant Video is the only U.S. online video subscription streaming service that offers offline viewing, and it's included at no additional cost.

X-Ray for Music

When Amazon introduced X-Ray for eBooks and textbooks, the world rejoiced, thanks to the ability to look up pertinent information about characters, places and events without leaving the current page they are on. Now, expanding its technology to other platforms, X-ray is available for music!

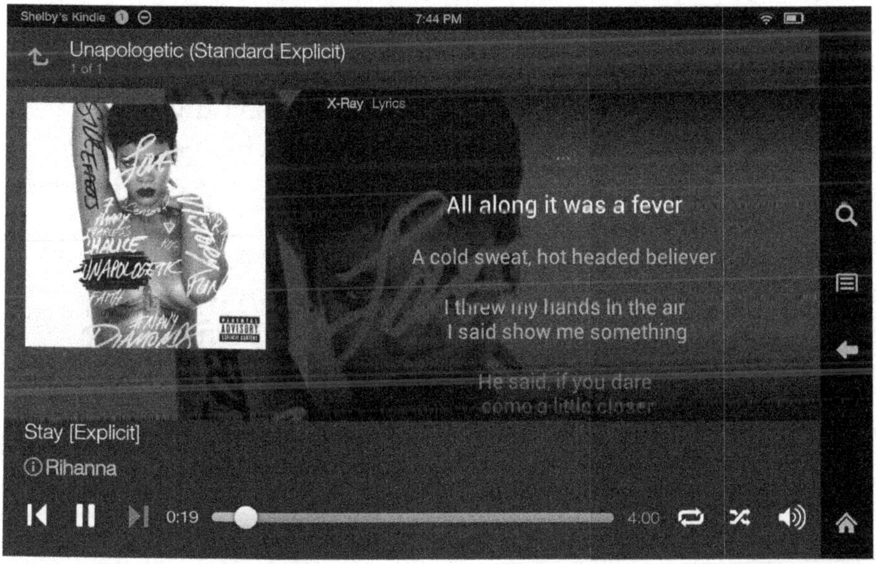

As a song in plays, X-Ray will list the artist and title, along with the lyrics. The lyrics will be displayed on the right side of the screen, and will scroll line by line as the song plays. You can see an example in the screenshot above. So now you never have to wonder what your favorite artist is saying, or look to another source for the information. It is all right in front of you with X-Ray for Music.

E-Mail and Calendar Support

With the email exchange ActiveSync, you can check your email accounts on the go, while responding accordingly and opening documents with ease. The emails will stay in your inbox until you delete them, so you can stay organized at the office, even while you are traveling.

In addition, when you schedule appointments in your email calendars at home or the office -from your laptop, smartphone or desktop - it will sync to each of your devices, including your Kindle Fire HDX. This means you will be alerted to a conference call in the morning, which may have slipped your mind, no matter which device is closest to you.

This capability is perfect for the busiest of bees that count on their KFHDX to remind them of everything from birthdays and anniversaries to meetings and doctor appointments.

SECOND Screen

Much like Apple has Airplay, which allows you to turn your iPad into a remote for your big screen with an Apple TV streaming device, Kindle Fire HDX has followed suit by turning your tablet into an entertainment hub that "Flings" content onto a larger screen, to view with a group. This content can be pictures, videos, movies, television programming or even your email inbox. All you need is the KFHDX, a Samsung television or a PlayStation 3 (or 4, when they arrive) for streaming purposes, and you are all set. The best part is, you can continue to work on your tablet while the television displays alternate content, so you can check emails or look for content on the web as you desire, while enjoying the programming remotely.

Miracast Capability

Very similar to the Second Screen capability, Miracast allows you to wirelessly "mirror" movies and television shows to your television screen (or any larger screen) that has an HDMI connection, simply by purchasing a dongle that is compatible, so the entire room is able to enjoy what is on your tablet's screen.

This can work for presentations and slide shows just as well as it can your favorite movies or videos!

Quick Switch

Quick Switch is the answer to all of your content woes, simply because it lets you toggle anywhere in the system to navigate between multiple apps - which is not uncommon on Android devices. What is new is that you can - with a sweeping single swipe - you can work within individual content so you can toggle between books or textbooks without navigating to the home page. You save time, and stay organized, without getting distracted by things that are happening elsewhere on the device, which means more work (or play) is getting accomplished.

Quiet Time

If you have ever read an eBook on an electronic device, you have suffered from an interruptive alert while enjoying what would be your quiet time. Things like sports updates, breaking news and even calendar reminders will all interject and appear on screen, no matter what you are doing. The KFHDX has remedied that by adding a "Quiet Time" button that prevents notifications from interrupting your reading time. It is listed as its own button, and is located right next to the Mayday button.

New Accessibility Features

The new accessibility features on this device are really helping its popularity, with items such as Screen Reader, Explore by Touch and Screen Magnifier. Each of the options are designed for the visually impaired to use the Kindle Fire HDX effectively.

The Screen Reader features the device's inner-voice - Ivona - and allows her natural language to dictate text to speech, in a voice that is less robotic than other tablets might include. This allows the device to explain what is happening on screen in a tempo that you prefer. You can slow down the translation or speed it up as you see fit using one of five speeds.

Explore by Touch verbally describes the items that are tapped on screen, so the proper apps or information can be accessed effortlessly. And the Screen Magnifier allows the visually impaired - or even those who have misplaced their glasses - to zoom in quickly on any material in front of them, including the home screen.

HTML 5 Apps Support

The Kindle Fire HDX also ships with built-in support for Google Apps for Business, for users who prefer Google's cloud-based applications to OfficeSuite. In addition, the company is talking up support for HTML5 apps and the ease with which developers can port titles over from Android. This will allow users to expand their apps options as they see fit, for personal and business use, do your device is never lacking.

Complete Enterprise Support

When you begin grouping Cloud Collections, office documents and email threads, or even printing wirelessly, these enhanced enterprise offerings are completed supported by this single device, so you are never at a loss for professional storage, compatibility or even crossover necessities.

Optimizing Download Manager

The new Optimizing Download Manager enables parallel content downloads. Unlike standard Android, Fire OS adjusts the number of simultaneous downloads per device based on the bandwidth, so that foreground app performance is not impacted by background download activity.

In addition, the Optimized Download Manager automatically pauses ongoing content downloads when the customer starts streaming an Amazon Instant Video, to maximize the video quality. Customers can prioritize individual items to download first, and with progressive download they can open videos, Audible audiobooks, or large periodical files and start enjoying them while the download is still ongoing.

Features No Longer on the Kindle Fire HDX and HD

With the redesign to the Kindle Fire HD and the new Kindle Fire HDX, Amazon did away with a few features that users may expect or be used to having with their tablets.

No HDMI Port

Both the Kindle Fire HD and Kindle Fire HDX no longer have the micro HDMI port, which allowed users to easily connect to HD television sets. While this is causing many users to be upset, there are a few workarounds to still be able to connect your tablet easily to your HD TV set.

One thing you can try for this is to buy a Micro USB to HDMI adapter. This may allow you to connect the wire to your TV's HDMI slot and play videos and slideshows on your big screen TV.

However, if you do not want to mess with the wire, you can use the Kindle Fire HDX's Miracast technology, you can use the Second Screen technology or try sideloading Google Chromecast (if you own the app files), to play various videos and slideshows on your TV.

No Camera on Kindle Fire HD

The basic Kindle Fire HD no longer has a camera. Some users may miss this option especially if they like to video chat or take self-pictures using their tablets. The Kindle Fire HD has no camera available, but the good news is, the camera was not super high quality anyway.

Even so, if having a camera is important to you, you will have to buy a Kindle Fire HDX model to get the camera.

Apps You CANNOT Use on KFHD and HDX

Since technology is still divided in it use, as it has been for decades, some apps are simply not compatible with the KFHDX. Just like Apple does not promote Flash, this device has a few apps that are not optimized for its use. It is their loss, really.

- Google Maps (or any maps): There is no GPS built into either version of the Kindle Fire HDX, so mapping is completely out of the question - from an app standpoint. Google earth or Google apps are also not part of the package.
- Instagram: All of those square, filtered pictures you have enjoyed up to date are going to have to come from your smartphone, because the KFHDX does not support the app.
- Google Chrome & Mozilla Firefox: The Kindle Fire HDX comes with "Silk" as its provided browser, and you will not be able to override it with a preference of your own.
- Google Drive: If you count on Google Drive to store documents you need throughout the day, you may want to consider switching to Amazon's Cloud for these purposes, as you will not be able to access your library from your Kindle.
- YouTube: You can still use their mobile site, but you will not have access to the app.

There are also a number of Android widgets that are not compatible with the Kindle Fires. Since the number of widgets available is practically innumerable - and the varying options that people use are almost as high in quantity - it would be impossible to list all of them here. Just keep in mind that some of your widgets may not make the cut, but you should be able to find an Amazon friendly alternative.

If you really want these apps, you can attempt to sideload them onto your Kindle device, per the instructions in this book, but keep in mind the legal issues involved with sideloading apps.

Other Helpful Kindle Fire HD and HDX Features

There are many interesting features available for the Kindle devices. Below are several of these features.

X-ray Features

The Kindle Fire HD has expanded its X-ray technology to include more than just books. Initially, the feature was available to allow readers to look up the meaning of a word by simply tapping on it, or to review the when, where, what and why of an event, place or occasion.

That technology has been developed for use during movies so you can check on an actor's name, biography, other roles and scenes without leaving the movie. It delivers the ever-popular IMDb app information within the movie. To access this feature on a movie that offers X-ray, all you need to do is tap the screen during a scene and it will pop-up information such as which actors and actresses are appearing and more!

In addition to regular books, X-ray is now available for textbooks, providing the user with Wikipedia and YouTube glossary references to enhance the learning experience, without leaving the page.

X-ray for music is also available for Kindle devices. The words to X-ray enabled songs will scroll on the right-hand side of the screen as you hear them.

Immersion Reading

There is a new way to enjoy books with this generation of the Kindle, and it is called Immersion Reading. When enabled, an audiobook will read the words, as the screen highlights them. This feature is perfect for beginning readers, allowing them to follow along within the text of the book, while the voiceover reads aloud.

Whispersync for Books, Games, Movies, and Voice

Whispersync for Voice is cutting edge technology that allows you to toggle between reading a Kindle book and listening to its companion audiobook without losing your place. It will also remember your position, and keep your notes and bookmarks.

When used for games, movies and books, this technology will sync each device you are enjoying the entertainment on, placing you directly where you left off on the other.

Cloud Storage vs. Device Storage

Amazon has made technology more user friendly than ever before by allowing you to store any Amazon purchased content on their Cloud, instead of devour the allotted storage on the actual device.

Depending on the version you purchase, you may be limited to 16GB of device storage, which means any non-Amazon purchased content will start eating away at that space immediately. This can include any iTunes music, videos, movies, or television content you purchased through Apple. However, each and every piece of content you purchase through Amazon can be stored on their Cloud server for free, without any size restrictions. This virtual storage capability can be accessed from anywhere with an internet connection, even if you do not have your Kindle Fire HD handy.

What You Can Store on Your Tablet

There is no limit to the digital files you can store on your tablet, including music, movies, apps, documents, contacts, eBooks, games, pictures and video. As long as it fits within your storage parameters, you can pack away files left and right until you are at capacity.

You can transfer images, content and technology to your Kindle Fire device from your laptop or desktop computer, camera, cell phone, or video camera and access it as often as you would like going forward. You can also transfer items from your tablet to USB jump drives, computers and storage devices effortlessly, as a backup precaution. Later on in this guidebook, there are instructions for moving files to and from your Kindle Fire HD and a PC or MAC computer.

Extra Storage Options

There are many free apps out there allowing you to store extra files online. These include Dropbox, which is a very popular app for file storage and sharing, but it will be limited to 2GB of storage for the free version. Google Drive will also let you store various files online for access on computers and devices. You may need to install an app to use Google Drive with your Kindle Fire HD.

Additionally, you can purchase special wireless devices to hold your files for access at home on a wireless network. The Kingston Wi-Drive is available at Amazon.com in 16GB, 32GB and 64GB sizes. It will hold photos, music, documents, videos, and other files, and will allow you to access them wirelessly on your Kindle Fire HD. These devices will allow other computers within your home or office network to share and stream the files in addition to your Kindle Fire HD.

There is also the Maxell AirStash Expandable Capacity Wireless Flash Drive, which is available in 6 gigabyte or 16 gigabyte sizes. It is more expensive than the Wi-Drive, but a smaller, more portable option for file sharing wirelessly on your Kindle, computers, and other devices. These items will include specific instructions to help you set them up easily with your home network and you can then use them to store and access files with the Kindle Fire HD.

Kindle Fire HDX Tips and Tricks

While you probably have the basics down, there are some neat Kindle Fire HDX tips and tricks to use to get the most out of your Amazon tablet.

Music and Videos

Although your KFHDX will not support iTunes, it will allow you to drop and drag your own music and videos onto the device effortlessly. Simply attached the device to your computer or laptop, and drag the files you wish to appear on the KFHDX accordingly. For music, the device supports MP3, AAC, FLAC and OGG filed. For video, the device supports H.264, MPEG4, Xvid and DivX videos up to 1080p.

In addition, Netflix and Pandora work beautifully on this device, which will allow you to stream unlimited movies, television and music with the appropriate subscriptions to each service. Both have a free trial, so you can experience their capabilities before settling into a monthly subscription service.

The high-resolution screen that is included with the KFHDX provides an exceptional viewing experience for HD movies, and really brings your device to life while watching any form of entertainment. The onscreen clarity and vividness of the colors is unbelievable, and will enhance your viewing experience each time a high definition option is available.

Just as X-Ray is available for music - to help you learn all of the proper lyrics to your favorite songs, it is also available for movies! This feature allows you to get information on the actors, or even the backstories characters, simply by tapping on the feature button. Now you will never miss a single detail of a new show, or even an old favorite.

What is Sideloading?

Sideloading is a fancy term for transferring media files from one device to another, usually from a USB drive, Bluetooth technology, or memory card. This ability allows you to transfer files effortlessly, which means you can have all of your images, documents, videos and music in one place, even if it is coming from someone else. Think of sideloading as something you have always done with your laptop or desktop computer, and expand that thought process to your new Kindle Fire HD. For the most part, the terminology here refers to a way to put apps onto your tablet that aren't necessarily available through the Amazon app store.

Sideloading Apps You Want for the KFHDX

It is possible to transfer some of your favorite Android apps from your current device and side load them onto your KFHDX. There are two options that make this possible, and you are free to use a combination of both.

Using the APK Extractor App, you can copy an app from your current Android phone or device, and attach it to an email. Email it to yourself, and download it to your KFHDX. Next, turn on "Apps from Unknown Sources" under Settings >Applications, and tap on the downloaded app. The app will install effortless, and function normally.

You can also install the free ES File Explorer app that is available through the Amazon App Store. Tap Quick Settings > More > Device, then tap "On" to Allow Installation of Applications from Unknown Sources.

1. Google your app and search for the APK file.
2. Download the APK file to your computer.
3. Copy the APK file to the root (or a folder of your choice) on the Fire via a USB connection to your computer. After copied, disconnect the Fire from the computer by clicking Disconnect on the bottom of the Fire's screen and then unplug the USB cable.
4. Start ES File Explorer and select the APK file then select install when prompted.

If you do not own another Android device, side loading an app is trickier because the only way to get a legal copy of an APK file is from an app you have downloaded from the Google Play Store or other Android app store.

How to use Send to Kindle Plugin for PC and Mac Web Browsers

The Docs area of your Kindle Fire HD or HDX tablet provides several ways to send documents over to your device. One of these ways is a feature you can use when online via a personal computer or laptop and you're browsing the web.

With the "Send to Kindle" feature, you can clip various web content such as articles or blog posts, and then have them automatically sent to your device to read later on.

1. You'll need to install a plugin for the web browser you use most, whether it is Google Chrome or Mozilla Firefox.
2. The plug-in should give you a small "K" icon for Kindle that shows up near the top of your web browser. You can click on this icon whenever you are on a webpage article you want to clip and send to your KFHD or KFHDX.
3. You'll be able to preview the item before it sends, and then click on "Send" from your web browser to send the formatted article to your device.

You can access all your "Send-to-Kindle" Docs by going to the "Docs" option on your Kindle's top menu and then tapping on the three lines up in the left corner of the screen to reveal a menu of options. Tap "My Send-to-Kindle Docs" to access your latest documents sent.

Note: *Keep in mind it may take several minutes for the document to arrive to your Kindle device. It will be sent via Wireless and could take some extra time to arrive. You can also try to speed up the delivery process on your Kindle tablet by tapping on "Settings" and then "Sync All Content."*

How to Opt out of Ads on Kindle

Some Kindle Fire HD owners may find the constant sponsored ads from Amazon on their device to be annoying. For a charge of $15, you can easily opt out of receiving the various sponsored ads from Amazon that show up on your Kindle Fire HD. You can do this by going to your Amazon.com account online and going to "Manage Your Device" to opt out of ads for the $15 charge.

How to Use the Kindle Fire as a Phone

A hidden gem of tips for your tablet is that you can turn it into a phone and make calls easily! Magic Jack has a free app available at the Amazon Appstore called "FREE Calls with magicJack." The beauty of this app is that you can install it for free on your Kindle Fire HD or HDX and then use it to make phone calls (although calls may be limited to Canada and U.S.).

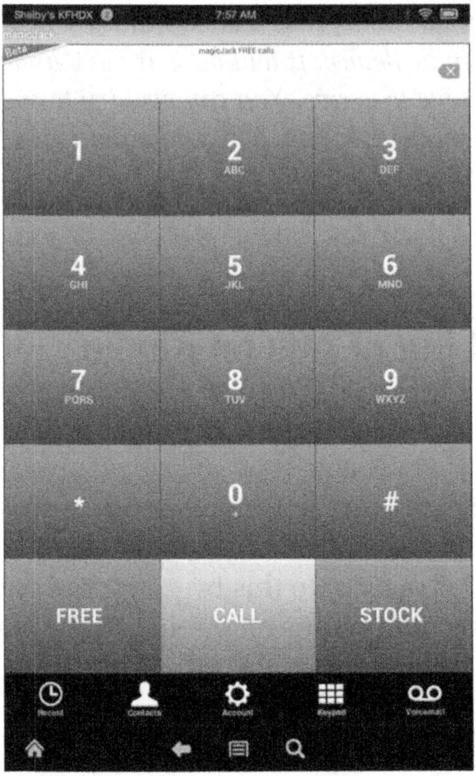

Once you install this app, you'll be asked if you have a Magic Jack account or not. You can tap on "Yes" or "No." Tapping on "No" won't require creation of an account and immediately brings up a screen with a large keyboard on it. There are also other options with this app such as adding contacts, setting up voicemail, and creating a Magic Jack account to use. The benefits to creating an account will be that you can create your own MagicJack phone number to use with your tablet.

Dial up a local number such as your home phone or someone's mobile phone to give it a try.

How to Conserve Battery Life

In addition to installing a good battery app such as Battery HD or GSam Battery Monitor, I recommend a few additional considerations and steps. These include shutting off Wi-Fi when you are not using the device with a wireless network, dimming the screen display brightness, and allowing the device to recharge fully when plugged in, rather than trying to use it while charging.

How to Clean the Kindle Fire Display

It is important to limit the amount of smudges and marks that get onto your Kindle Fire HD display so it will provide good viewing and functionality over its lifetime. The best investment here is the screen protector recommended earlier with accessories. If you don't have a screen protector, then a microfiber cloth is essential. With a microfiber cloth you can wipe off any smudges or dust to keep your display looking great.

How to Print from Kindle

While there isn't a built-in feature to print using your Kindle, you may consider several apps. These apps will vary in terms of functionality and feature, but the biggest factor will be your particular wireless printer. Some apps may not support some printers, and some printers simply might not work with the Kindle Fire HD device.

Among the print apps you can check out are EasyPrint and PrinterShare Mobile Print. EasyPrint is free and ad-supported app. It works using the Google Cloud Print service, which you will need to set up in addition to the app. There are more instructions on how to do this with the app and Google Cloud Print Service.

PrinterShare Mobile currently sells for a price of $12.95 at the Appstore. With this app you can print various documents wirelessly from the Kindle Fire HD to many types of printers including those from HP, Epson, Canon, Brother and Samsung, but you'll need to check for compatibility first.

How to Get More Kindle Support

Looking for more answers on how to use your Kindle device? The Amazon website provides plenty of extra help for Kindle owners. There is a variety of Kindle Self-Service tools at the site including:

Manage Your Kindle

Manage Your Subscriptions

1-Click Payment Settings

View Digital Purchases

View Your Collections

Kindle Help Forum

In particular, the Kindle Help Forum is a great resource for finding out answers to issues with your device that other members may have been able to resolve or address. There may also be new pieces of information from Amazon staff regarding future Kindle Fire HD updates there. To access these help areas and more visit the Kindle Support site.

How to get free eBooks for your Kindle Devices

There are plenty of free eBooks to be had for your new Kindle. Below is a list with links provided to different places you can visit online to find free content for your device.

- Amazon's Kindle Store. On the right side of the page, there is a list of the top Paid and Free books for the day. Click on that list, and you will have access to the top free books of the day. http://www.amazon.com/gp/bestsellers/digital-text
- Amazon Kindle Owners Lending Library (for Amazon Prime Subscribers) http://www.amazon.com/gp/feature.html?ie=UTF8&docId=1000739811
- Free Book Feed - http://www.freebookfeed.com
- Many Books - http://manybooks.net/

- Add All eBooks - http://ebooks.addall.com/amazonfree.html
- Bargain eBook Hunter - http://bargainebookhunter.com/category/free/
- Books on the Knob - http://blog.booksontheknob.org/
- Daily Cheap Reads - http://dailycheapreads.com/category/free/
- Daily Free Books - http://www.dailyfreebooks.com/
- e-Reader Café - http://www.thee-Readercafe.com/
- e-Reader Perks - http://www.e-Readerperks.com/
- Free Book Dude - http://www.freebookdude.com/search/label/Kindle?max-results=5
- Pixel Scroll - http://pixelscroll.com/category/ebooks/free-kindle-books/
- Your Daily eBooks - http://www.yourdailyebooks.com/category/free-kindle-books/

Helpful Tip: *You may also want to check with your local library. Many libraries currently offer eBook check out privileges for patrons to borrow various eBooks on Kindle devices.*

How to Root a Kindle Fire Tablet

Rooting a Kindle Fire HD tablet is not something recommended for most users, so consider this an advanced tip and with a word of caution: rooting your tablet will void the warranty, and it may also cause glitches or issues if done improperly.

However, doing this in the past has allowed some users to access much more content and make certain adjustments to the tablet that couldn't be done before. For example, some users have been able to install the Google Play store on a KFHD tablet, allowing them a greater selection of apps and content to use on their tablet.

To root a Kindle Fire HD or HDX tablet you'll want to check out forums online such as XDA Developers, where users will discuss ways to do this. Normally it will involve downloading a large batch of .apk files and then installing them with the free ES File Explorer app from the Amazon App Store.

Once again, only root your device if you have a good idea what you're doing, and can handle the potential issues involved.

Kindle Fire HD and HDX Parental Controls

As any parent is aware, technology can easily take over your child's ability to interact with others, or to even do their homework, if certain boundaries are not established. Even when they are understood, if you happen to be out of sight, or away from the house, the boundaries may be overlooked to steal away a few extra minutes of internet surfing. That is why the Kindle Fire HD provides Parental Controls that allow you to establish time limits for certain users, while enabling content controls for curious eyes.

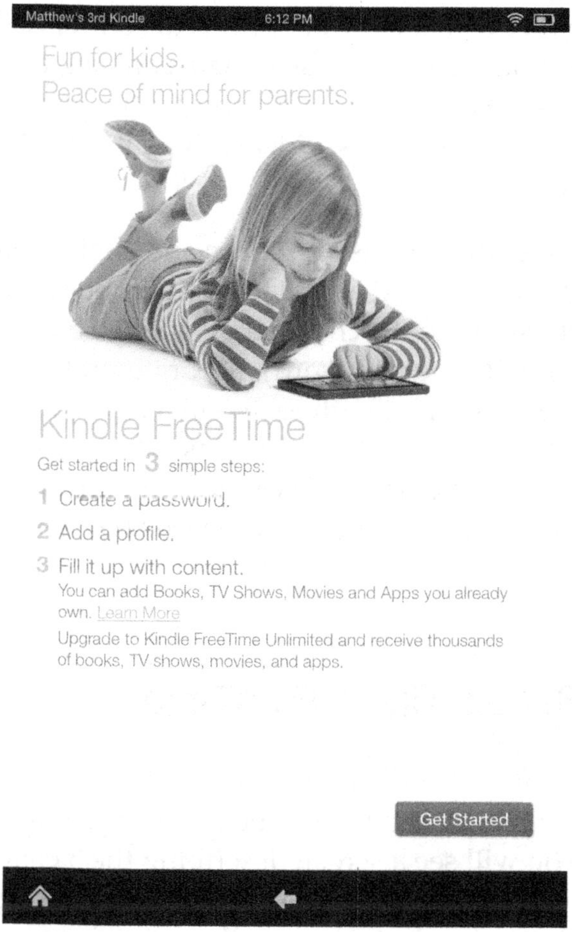

Kindle Fire HD provides parents with all the tools they need to police their child's device usage. In fact, it even comes with its very own name: FreeTime.

Swipe the top of the screen from the time display to list the settings menu. Tap "More" to display the settings in their entirety. Tap "Parental Controls" and tap "On" to enable the setting. Create a password, preferably one your child will not guess, and enter it twice to move forward.

You will be met with a bevy of options to control with your device including web browsing, e-mail, contacts, and calendars. In addition, and more importantly, you can password protect purchases, videos, content, and Wi-Fi connections.

You can also set time limits for games and content, but can give your child unlimited access to books or educational material, which cannot be changed by anyone but you (or another adult with the password). Separate profiles can be set up for multiple children, with each only having access to the material you, as the parent, establish as safe and relevant to their age and lifestyle. This will create a fun, safe and easy-to-use environment for younger Kindle Fire HD users.

How to Set-up Kindle FreeTime

1. On your Kindle home screen, tap the Apps in the top menu, and tap on the Kindle FreeTime app.
2. You will see a screen describing the 3 easy steps to set up your account. Tap on "Next" to begin.
3. Create a unique password you will remember that is at least 4 characters. (Make sure it is one that the younger users won't guess easily. You will need this password in order to unlock the device from FreeTime or change settings).
4. Confirm the password by entering it again, and then tap "Next" or "OK" to get to the next step.
5. Create a profile for a child who will be using the Kindle Fire HD. You can enter their first name, specify "boy" or "girl," enter birth date and add a photo if you want.

6. Once you have finished entering the child's profile info, you can tap "Add Another Child" to add more children who will be using the Kindle Fire HD, or tap "Next" to go to the next step.
7. The Kindle Fire HD will add the child user to your household as part of the Kindle device.

You'll now see any profiles for children you created, as well as additional options below where you can manage the content they access, set daily time limits for use of the Kindle Fire HD, and Manage Child profiles.

When setting time limits, you can set a total screen time that the child is able to use the Kindle Fire HD for. Or you can set specific "Content Activity Time" for reading books, watching videos, and using apps.

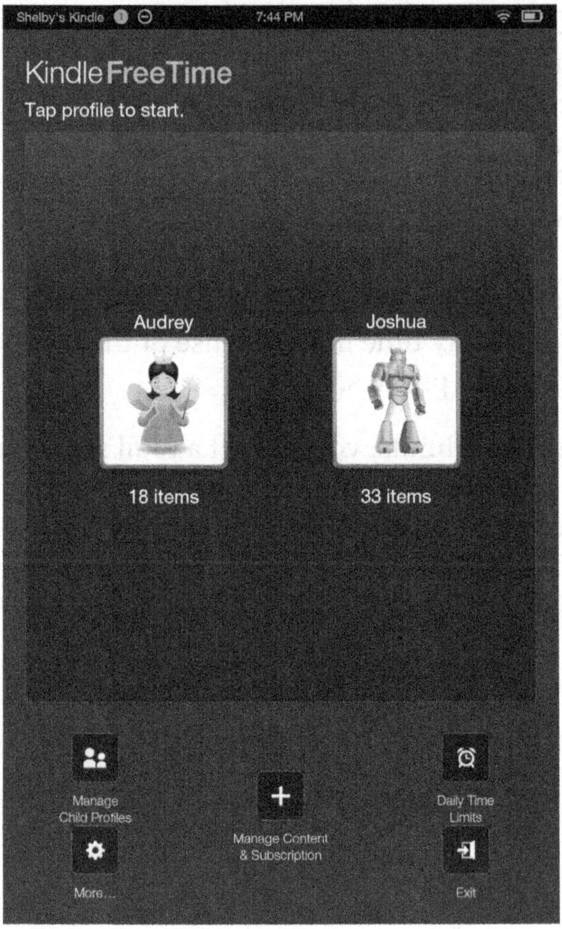

By tapping on the created profile photo (in the above image, the princess or robot) you will then load FreeTime for that particular user. The user can then choose from any of the pre-loaded content, or content you have purchased and placed there for them to access.

The user can use the Kindle Fire HD as you've set it up for them, but they will not be able to access your content, such as your apps, music, videos, movies, etc.

Exit FreeTime or Change Settings from Inside

To exit FreeTime or change settings from within an app, book or other piece of content, you'll need to tap on the screen near the top to drag down the notifications menu bar. Choose "Exit FreeTime" or the option you want to control, and enter your password.

Recommended Apps for Kindle Fire HD and HDX Owners

There are thousands of apps out there, but I recommend the following ones as some of the best basic apps you can get on your device. These cover the categories of news, productivity, utilities, and much more to help upgrade your tablet! The majority of these are available from the Amazon Appstore, but some are available from third party sites. We have covered these particular apps earlier in this eBook.

Candy Crush Saga - Finally this incredibly popular game from King is available for the Kindle Fire! Hooray for Candy Crush lovers everywhere. You no longer have to sideload this app onto your Kindle. Now you can crush candy to heart's content on your Kindle Fire devices simply by downloading this game for free from the Amazon app store. Be sure you are careful about how much you spend in in app purchases.

Angry Birds Free - This is the go to game that most people love to play. It is a great time-killer to have on your Kindle Fire HD, and great for the kids to play. It features a slingshot you use to fling birds across the screen in an attempt to destroy various structures and enemies for points. A very addictive and fun game, so be careful! There's also Angry Birds Seasons Free and Angry Birds Space Free for those who want to try different versions of the game.

Battery HD – This is a great app for monitoring the levels of your Kindle Fire HD's remaining battery power. It will give you various displays including info on how much time you have left for viewing movies or videos, browsing the internet with Wi-Fi, reading books or more. A great app to get to make sure you conserve battery power and know when to charge back up! An alternate battery app to check out is called GSam Battery Monitor, which will tell you which apps might be using the most battery power on your device.

Calculator Plus Free – The Kindle Fire HD does not include a calculator as part of its features, although many other tablets do. Calculator Plus Free is currently the most reviewed, and highest rated of the free calculator apps at the Amazon Appstore. An alternate app to this is called Calculator and costs 0.99 cents.

Dolphin Browser – This was mentioned previously, as it can be a much, more comfortable way than Amazon Silk to browse the internet. It also has been reported to allow the use of Adobe Flash, but there are other steps involved to set this up. You will need to install Dolphin Browser 8.5.1 to be able to use Flash.

Note: *Adobe Flash can be obtained at the XDA Developers forum. The download is approximately 4.5 megabytes in size.*

Dropbox or SugarSync – These apps offer you more cloud storage than what you get with the free Amazon storage. They also allow for easier movement, streaming and sharing of files between your devices.

ES File Explorer - This is hands down one of the best apps to install on your Kindle Fire HD to really unlock and unleash more capabilities. Among those capabilities is the camera functionality we covered above, as well as the ability to manage all the files and folders on your device.

Facebook app - It is tough to find the Facebook app at the Amazon App Store, however you can access it at the link on the app name here.

Hulu Plus - Hulu Plus is an online subscription service that allows its users to enjoy on demand access to television shows, movies, webisodes, news, trailers, clips, and behind the scenes footage for free from networks and studios, thanks to ad support. Many people opt for this service as an alternative to Netflix due to its variety of available television shows, movies and other entertainment.

imo Instant Messenger - Use this free app to connect with friends, family and colleagues across multiple instant messenger accounts. The app will worth with AIM, MSN, Yahoo, Skype, and much more!

OfficeSuite Professional 6 - While QuickOffice is a nice app for $15, there is also a free app called OfficeSuite Professional 6. This app is compatible with MS office files. It also allows you create, edit, view or save the various Office files, meaning more productivity on the road when you don't have access to your computer or laptop.

Netflix - If you currently have a paid Netflix account to enjoy streaming movies, you'll want to get this app. This will allow you to watch all of the same streaming Netflix movies you can watch on compatible Blu-ray players, computers, and other internet devices. A must-have for those with the Netflix subscription and a Kindle Fire HD who want to enjoy more movies than what Amazon has.

PicShop Lite Photo Editor - An alternate option in terms of camera apps is called PicShop Lite Photo Editor, also available at Amazon App Store. There are both free and paid versions of this app. The free version is very similar to the camera app discussed earlier in this guidebook, which can be accessed with ES File Explorer.

Pandora - Pandora is an alternate music app to Spotify (mentioned below). It can be used to play music based on radio stations created from your personal music preferences. For example, choose the artist Barbara Streisand, or Guns N' Roses, and it will create a radio station with music from the same genres as the original artist. This is a great free or paid app for streaming music radio style.

SplashTop Remote Desktop - Use this paid app to control your desktop from a remote location, such as while away on travel. Your desktop or laptop computer will need to be on to do so, and you'll have to install the SplashTop software on the device you want to control. Still, for $4.99 this is a great item for many people who want access to their home computer while on the road, or even from inside their home.

Spotify - Spotify is an online music streaming service that can be accessed by anyone who has a FaceBook account. Once you register your account, you can listen to unlimited music for six months, thanks to its radio style sponsored advertising. Music can be browsed by artist, song title, album title, genre or playlist. After the trial period, there is a ten-hour listening limit per month, divided into two and one half hours per week for unpaid subscriptions. An unlimited subscription is available, and provides access to music without advertisements or time limits. A premium subscription is available that provides the unlimited access, no commercials or time limits and a higher bit rate of streaming, combined with offline access and a mobile app accessory.

textPlus - Created by GogiiInc, this app allows you to send texts for free to people around the world who also have the textPlus app. Not only that, it will work for sending out text messages to people in Canada and the United States who don't have the app. The textPlus app is available in both free and paid versions.

TubeMate - A third party app not in the Amazon appstore, this will allow you to watch and download YouTube videos on your Kindle Fire HD. A great way to get videos to watch at a later time, such as on a flight or during travel, so you don't have to worry about having a wireless connection to watch them! This book provided detailed instructions in a previous section to help you get going with this one.

Tune In Radio - A great free app that allows you to stream local and national radio stations on your Kindle Fire HD. Listen to your favorite local radio station while doing other activities on the Kindle. It also features other categories such as sports, talk, news, and podcast programs to select from.

As noted, there are tons and tons of apps being sold with new apps created daily. Those listed above are just several options you can consider to really upgrade your Kindle Fire HD for entertainment, productivity and more. For even more great apps I recommend the Kindle eBook, "Best 100 Kindle Fire HD Apps," which covers a plethora of great apps, both free and paid.

Accessories to Consider for KFHD and HDX

As with any great device, there will be a number of accessories available to highlight its attributes, while you enjoy the positive and complementary functionality that they deliver. No matter who you are, or your personality, there are accessories available that will reflect your need without an issue.

Cases

Available with the Kindle Fire HDX devices is the Origami case that folds in a savvy design, to bring a little flair to your device, while propping it up on the backside, you do not have to hold it during a movie or presentation.

In addition, there are leather cases, and Otterbox "Defense" cases - that will envelope your device into safety so even an inadvertent drop will not harm your favorite device. Also, for the daintier owner, there are design emblazoned slim sleeve cases, so you can carry your Kindle Fire device in style.

Stylus

Sometimes it just makes more sense to have a "pen" when you are operating your device, especially if you are in a meeting or presentation during its use. Purchasing a stylus pen to control the device, while tapping apps, adding content and viewing material is a mere $10 away (or less).

If you choose to turn your entire tablet into an interactive device, a Stylus Suite is available that allows you to hold the device with a built in elastic strap, and includes a built-in i-Blason Stylus/Pen Combo. The latter magnetically closes and includes Built-in ID and Credit Card, and SD Card Holders for your convenience.

Screen Protector

Outside of a case for your little beauty, a screen protector is probably the next most important accessory you can purchase. First, it keeps any inadvertent scratches from appearing on the screen, should you happen to slide it into your purse, briefcase or backpack for a quick transport. Likewise, you never know when something sharp or pointed will come in contact with your device, so it is better to be safe than sorry.

In addition to it protecting the screen from harm, it is also cuts down on fingerprints and glare, so you can enjoy your device at any time, without feeling like it needs a good wipe down.

Miracast Device

In order to mirror your device's content onto an HDTV screen (television, monitor or otherwise), you will need a Miracast Wireless HDMI Display Adaptor like the NETGEAR Push2TV Wireless Display HDMI Adapter. This device will cast - or "mirror" you screen's content directly onto a larger screen wirelessly. Its incredibly portable size reflects that of a jump, data drive, so you can carry it effortlessly and use it whenever a meeting breaks out off site.

HDMI Adapter

Since the latest KFHDX devices have dropped the HDMI port as part of their redesign, purchasing one will become a necessity if you still need to connect to an HDMI port manually. These adapters are available in micro-HDMI, so you can connect to your existing television and watch movies, sports, television shows or simply share your content with a room full of people, without worrying about wireless interference.

Samsung TV or PS3/PS4

In order to use the new, super cool Second Screen option that is available on the KFHDX, you are going to need a streaming device that makes it possible. Samsung Smart televisions and the PlayStation 3 or 4 will provide that access, so you can screen movies on the television for your family's enjoyment from your KFHDX, and still catch up on emails or change your fantasy football lineup without interrupting the programming.

Bluetooth Keyboard

One large complaint about tablet devices is that using the onscreen keyboard can give you a complete headache - while making your feel like you have the largest fingers to every maneuver a device. It just is not as easy as it appears, unless you go into "search and peck" mode while using the keyboard. For this reason, wireless, Bluetooth keyboards have become incredibly popular for the KFHDX.

You can purchase the Bluetooth keyboard to stand alone or as part of the travel/protective case, depending on how often you use the accessory. It is perfect for typing long emails, reports or even creating presentations on the go, so you receive the benefits of a real-sized keyboard, with the functionality of a tablet.

Bluetooth Headphones

Much like your regular headphones or ear buds, Bluetooth headphones allow you to enjoy the sound of music, movies or audio books, without physically connecting the headphones themselves into the device. This is great if you would like to keep your device tucked away, or do not want to be tethered to it as you sit at your desk, perhaps, so there is no chance of you sliding your chair across your office, and dangerously dragging the device with you. This is perfect for at home use too, if you are working around the house or garden, cleaning or preparing meals, as you can listen to your favorite audio items while you move about freely.

Headphones

If you are sitting still, and have the device in front of you, say on an airplane or train while watching a movie, your regular headphones will attached to the device with ease, allowing you to listen without interruption. At the current time, there are a lot more standard headphone varieties available on the market than their Bluetooth counterparts, but give it time. They will eventually catch up!

Bluetooth Speakers

Much like the Bluetooth headphones, a wireless, and Bluetooth speaker will allow you to connect your music to a speaker that can be moved anywhere in your home or outside, so you can entertain literally wherever you are. You can take it to the beach, to a picnic, or simply allow it to ride with you on road trips, so you can access it in your hotel room. This incredibly smart device works the same way all Bluetooth capable devices do. All you have to do is turn it on and allow your KFHDX to recognize it, before delving into your entire musical library or Pandora, so you are never without tunes - no matter where you are.

Kindle Fire Conclusion

The Kindle Fire HDX contains a number of new features and exciting options for tablet users. Not only is a lightweight and beautiful device, but also its functionality has surpassed a number of tablets on the market, and is available at a lesser price point. Once you see the amazing screen, and how the colors come to life during movies and videos, you will wonder how you ever enjoyed your favorite television shows before. The same goes for the images you take with the device, and the ones you manually transfer. The Cloud capabilities outshine its competitors by providing 1-Tap archiving, and easy to access folders and files that are intuitively added as you see fit.

The whole device allows you to access its features without getting a headache, as the flow of its design, and its amazing functionality will walk you through the lineaments without issue. Should you get stuck, anywhere - at any time - simply tap your Mayday button and wait for onscreen help to arrive. Remember, this option is available 365 days per year, 24 hours per day, so help is only a tap away. There is no question too small, and none too large for the customer service representative to walk you through.

Finally, the apps and capabilities of the Kindle Fire HDX will allow you to combine professional and personal interests on the same device, without worrying about your company's ability to interfere with the segment that does not belong to them. Everything that is downloaded to your KFHDX is protected, and accessed only by you, so Human Resources is not going to tap into your personal Hotmail account, simply because your office account is available on the same device.

In a world of advancing technology, tablets are going to continue to upgrade and become newer with each step of the process. So far, the evolution of the Kindle devices that led up to the HDX has nailed every segment of the market, allowing their clients to enjoy the benefits of Amazon actually listening to their feedback from previous iterations of the same device.

Take the time to discover which part of the KFHDX works for you, and the only decision you will have to make after that is whether to go with the 7" screen or the 8.9" screen. The choice is yours, and so are the amazing accessory options that will allow you to turn your Kindle Fire HDX into a personal expression of yourself - by day or night. All you need is a quick case change to take you from business to creative in a matter of seconds, so enjoy the versatility and the design at every turn. That is exactly what they are there for!

Thank You - Free Kindle Fire HD and HDX Apps eBook by Shelby Johnson

As a thank you for purchasing my book, I am providing you with another eBook I have written, free of charge! The Kindle Fire HD apps guide includes over 100 apps I have personally selected for owners of the tablet to install on their device. Many are free apps which you can add to your Kindle Fire HD today.

You can download the free Kindle Fire HD and HDX apps guide at http://techmediasource.com.

More Books by Shelby Johnson

iPad Mini User's Guide: Simple Tips and Tricks to Unleash the Power of your Tablet!

iPhone 5 (5C & 5S) User's Manual: Tips and Tricks to Unleash the Power of Your Smartphone! (includes iOS 7)

Facebook for Beginners: Navigating the Social Network

Kindle Paperwhite User's Manual: Guide to Enjoying your E-reader!

How to Get Rid of Cable TV & Save Money: Watch Digital TV & Live Stream Online Media

Chromecast Dongle User Manual: Guide to Stream to Your TV (w/Extra Tips & Tricks!)

Samsung Galaxy S4 User Manual: Tips & Tricks Guide for Your Phone!

More Kindle Fire HD Books

More Kindle Fire HD eBooks:

A Newbies Guide to Kindle Fire HD

Best 100 Kindle Fire HD Apps

Help Me! Guide to the Kindle Fire HD

Kindle Fire HD GuideBook

The Kindle Fire HD Manual

The Ultimate Kindle Fire HD GuideBook

Made in the USA
Monee, IL
04 November 2026

67384441R00069